Ocular Motility

Virginia Carlson Hansen, CO, COMT

SLACK Incorporated, 6900 Grove Road, Thorofare, New Jersey 08086

Managing Editor: Amy E. Drummond
Production Manager: John H. Bond
Cover Design: Linda Baker
Publisher: Harry C. Benson

Printed in the United States of America
Library of Congress Catalog Card Number: 87-42953
ISBN: 1-55642-028-5

Published by: SLACK Incorporated
             6900 Grove Road
             Thorofare, NJ 08086-9447

Last digit is print number: 10   9   8   7   6   5   4   3

About This Series

The field of Ophthalmic Medical Assisting has been developing since the early 1960s. As new technology was introduced into the field, technical tasks were delegated to office personnel in order to provide the physician with more time for a high-tech examination and treatment role. Each year a greater number of these advanced technology skills filter down to technical personnel, who need to be competently trained. In order to meet this demand for education, various formal educational programs for the Ophthalmic Medical Assistant have been developed; and to date ten programs in the United States are accredited by the American Medical Association. Various organizations participating in the Joint Commission on Allied Health Personnel in Ophthalmology have collaborated to define the three technical levels (assistant, technician, technologist) and continued to update the criteria for certification of each level.

With the goals of quality care for the patient and professional enjoyment for the ophthalmic team, we will approach the turn of the twenty-first century more successfully with education. All members of the ophthalmic team constantly need to sharpen their clinical expertise, acumen, and awareness in order to identify serious pathology outside the daily routine of the practice. This attention to detail enhances the quality of care we want for our patients and the professional confidence gained from a job well done. It requires a broad education, which includes a general understanding of anatomy, disease, diagnosis, and management, and a set of ready references to remind us of key concepts and methods. This series is designed to update and supplement your current knowledge. It is a reference guide to help the beginner as well as the experienced practitioner and those reviewing for their certification.

Susan C. Benes, MD
Department of Ophthalmology
Ohio State University

Contents

Acknowledgments

I'm tempted to say that I don't need to thank anyone because I did this all myself, but it just isn't true. I had plenty of support and encouragement.

I was very happy to be asked to write the Ocular Motility book for the Ophthalmic Technical Skills Series and appreciated the enthusiasm of the people around me. Thank you to all of them. Norma Garber, CO, COMT, helped get me into this and I suppose that I can thank her now that it's over with. (In the middle of the writing of Chapter 3, "thank you" was not the expression I would have directed toward her!) The Department of Ophthalmology at New England Medical Center provided both the tools and financial support needed to write this book. Dr. Jeff Brenner of NEMC helped me out numerous times with word processing problems and draft printing. Our department photographer, Jennifer Hamlin, made the photographic detail almost painless; and Dr. Anne Moskowitz straightened out my thought processes more than once. My father kept my subjunctive mode correct, and my mother told me when I was sixteen years old that ophthalmology was a good field to get involved in. (She was right: people will always have eyes.) Thank you, Scott, for being patient and for keeping dinner warm on those frequent late nights (and for proofreading the whole works); and thank you, Montana, for still wagging your tail when I got home late. Most of all, thank you to the little one who has been with me since I agreed to write the book and kept my mind on better things.

Ginny Hansen

About The Author

Ginny Hansen is a certified orthoptist and certified ophthalmic medical technologist. After training at the University of Florida in Gainesville, she worked for an orthoptic clinic sponsored by the Lions Organization and also for a private ophthalmologist in Massachusetts. During the 1980s Mrs. Hansen was the program director of the Tufts Orthoptic Program and in 1985 was appointed as an instructor in Ophthalmology by the Tufts University School of Medicine.

Throughout her professional career, she has been active in the American Association of Certified Orthoptists, serving on its executive board and board of directors as vice president, American Orthoptic Council representative, secretary, and eastern regional representative. In 1983 she received the honor certificate from the AACO and AOC. She is a member of the American Association of Pediatric Ophthalmologists and Strabismologists.

Since moving to Minnesota in 1990, Mrs. Hansen works in a private pediatric ophthalmology and adult strabismus office at Park Nicollet Medical Center in Minneapolis. It is one of the largest private group practices in the United States with over 300 physicians of all specialties. Because of the unique workplace setting, the opportunities for research in a primary care referral base of patients, is also unique. Mrs. Hansen is currently working with her pediatric ophthalmology colleagues setting up a computerized database collection system that will enable them to study trends in pediatric ophthalmology patients and adult strabismus.

Foreword

Strabismus! The very term evokes from beginning orthoptic students a passionate out-pouring of emotions, most of them unpleasant and suited more to the Inquisition than to present realities. One common view of strabismus is that of a catacomb filled with deep and dark secrets understood by only a select few.

Why this antipathy and dread of learning strabismus? One reason is that the strabismus evaluation is typically performed on a child with a vexingly brief attention span. The instructor simply does not have the luxury of repeating a clinical finding again and again until it is appreciated and can be elicited by the student.

As if constraints of the clinical examination upon learning were not enough, the raw material to be digested by the student is considerable. Consider the complex permutations of primary, secondary, and tertiary actions of the extraocular muscles, primary and secondary deviations in restrictive and paretic strabismus, anomalous innervations and other mechanical aberrations. Now add to these the protean sensory adaptations these motor abnormalities may provoke in the immature visual system, and one begins to appreciate the amount of material to be assimilated.

One solution to the problems of learning strabismus would be to provide *readable* text. I am speaking of a text that does not presume a generous background in medicine and anatomy, and one that does not burden the reader with minutiae that are of limited interest and importance. The text should be written so that were I still a resident, even I could understand it (which means that *anyone* can understand it)! It should be a text that shatters and illuminates the deep and dark secrets of strabismus. In other words, *it should make sense!*

I am pleased that *Ocular Motility* is just such a book, an introductory text to strabismus that is understandable! Despite its brevity, all the important concepts of strabismus are here. Strewn among these pages is a wealth of clinical pearls acquired by the author through many years of managing strabismic patients in her own practice, and through teaching a generation of orthoptic students and ophthalmology residents. My only regret is that this particular trove of insider's information did not exist when I began my own adventure into this exciting field.

Paul D. Reese, M.D.
Director, Pediatric Ophthalmology
Tufts—New England Medical Center
Boston, Massachusetts

Preface

Ocular motility is a difficult subspecialty of ophthalmology. It is often made more difficult by the nature of the typical patient — young, uncooperative, nonverbal, but having a good sense of how to dismantle the office. It is my hope that this book will change your opinion about ocular motility and transform its frustrations into challenges. Although I can only touch the surface of a very complex subject in this brief text, I hope I will enable the reader to understand the basics of an ocular motility exam and how to approach the patient from a practical viewpoint. Of most significance is the chapter outlining a systematic approach to strabismus. Because the majority of strabismic patients are young children, they must be approached as though each piece of information gained may be the last. The systematic approach to strabismus will enable you to organize what is known so that it can be used most effectively.

Introduction

This book is intended to change your opinion about ocular motility, a difficult subspecialty of ophthalmology. Its problems are compounded because most patients are young children—uncooperative, squirming, and plotting how to dismantle your office. I hope to change your opinion about the field by making it a challenge rather than a chore. Being a "good examiner" of a patient with an ocular motility problem requires two things: that you perform the tasks accurately, and more important, that you know why the exam is being done in a certain way using certain steps. Once you understand the "good examiner's" motives for doing particular tests, you yourself will become that "good examiner."

I want this book to show you that there are methods to the seeming chaos of an ocular motility exam. When used correctly, these methods help lead to a correct diagnosis, appropriate treatment, and a realistic prognosis. Ocular motility testing is a challenge with a good reward.

The chapters are presented here in a particular order for a reason. The initial chapters cover basic information that will equip you for what follows. The first areas cover some basic knowledge about muscle anatomy and function, binocular vision, and the various classifications of each type of strabismus. The adaptations to strabismus and the specific tests for them will follow along with history taking. Once these initial chapters are understood, the systematic approach to taking measurements will make sense. Ophthalmic medical assistants (OMAs) are more effective when they understand the methods used to make a diagnosis. A chapter on learning to use a differential diagnosis deserves space in this book along with the different modes of possible treatment that are available.

The final part of any ocular motility exam is explaining the short and long term prognosis to the parents or patient with an ocular motility problem. Those people need it explained succinctly, in terms that they can understand. It's the OMA, the nurse, and the therapist who have the time *and* knowledge to explain the diagnosis, treatment and prognosis well. Patient compliance increases with patient understanding, so the OMA's role is not finished until the patient fully understands the problem.

As you read and understand the chapters of this book, my best advice is to then practice what you've learned on a normal cooperative adult. Then practice on a cooperative adult with strabismus. Then find a fairly uncooperative adult with strabismus and try your techniques on them. Only then are you prepared to examine a child. I'm certain that you can find an uncooperative one to give you a really good test of your new skills. This will be the test to determine if you've really become a "good examiner."

CHAPTER 1

Extraocular Muscles: Anatomy and Function

Don't memorize extraocular muscle (EOM) function; it's not worth the agony or the imperfection. If you understand a few anatomical locations and remember those, the function of the muscles while the eyes are in various positions will be easy to remember.

Eye Movements

The sophistication of our EOMs allows for many different kinds of eye movements. Moving the eyes enables the field of view to increase and allows the fovea, the small region of good vision, to move. Our eyes can involuntarily follow a slow moving object (**smooth pursuit**), or they voluntarily jump from one fixation to a new object of regard (**saccade**). Once we are viewing the new object, our eyes maintain fixation by constantly readjusting this fixation so that the object remains on the fovea (**microsaccades**). As our heads change position in space, information from the balancing apparatus in the inner ear sends messages to the EOMs to "right" the globes (**vestibular**). Our EOMs also use information from the "position sense" mechanisms (**proprioception**). All these eye movements can occur monocularly. **Vergence movements** allow *both* foveas to maintain fixation at the same time (**bifoveal fixation**), even as an object moves closer (**convergence**), or further away (**divergence**). The rewards of maintaining bifoveal fixation are **binocular vision** and **stereopsis**. We can appreciate true depth perception. (Gay et al., 1974). So how do the individual EOMs work to accomplish all this?

Say you're about to step off the curb. Suddenly, you stop and look to the left in time to see that taxi about to whisk you off your feet, but not in a romantic way ... What do your eyes do? First, your brain received messages: you may have seen the taxi moving way out in the periphery of your

Eye Movements:
1. Smooth pursuit
2. Saccade
3. Microsaccades
4. Vestibular generated
5. Proprioceptive
6. Vergence movements
   Convergence
   Divergence

retina, or you may have heard it, or smelled it, or felt the wind from it. Or perhaps you've learned about that particularly dangerous intersection. Something alerted your brain, which sent messages to the eye muscles telling them to move. It also told your body to stop, and to tense up with adrenalin. There are nerves from your **brain stem**, the part below the thinking cerebrum which initiated your reaction, that can command eye movements. These special nerves are the **cranial nerves** (CN).

Neurological Basis of Eye Movements

The twelve cranial nerves are responsible for various duties, both motor and sensory, that help with basic functions (see Table 1.1). Many texts specifically outline these nerve pathways from the brainstem to the muscles around the eyes (Bajandas, 1980). This anatomy in the brain is important because it sets the scene for many EOM syndromes and locates neurological diseases that affect the eye muscles in a specific but unusual way.

Bony Orbit

Orbital Bones:
1. Frontal
2. Lacrimal
3. Maxillary
4. Zygoma
5. Ethmoid
6. Sphenoid
7. Palatine

The human orbit's lateral and medial walls form a 45° angle where the medial walls of each eye are parallel to each other, and the lateral walls of each eye from a 90° angle to each other. The superior orbital fissure is at the back of the orbit. Nerves to the EOMs and eye enter the orbit through the superior orbital fissure (see Ophthalmic Technical Skill Series — *General Medical Knowledge*).

Table 1.1 Nature and Function of Cranial Nerves

Cranial Nerves	Sensory Motor	Function
1. Olfactory	Sensory	Smell
2. Optic	Sensory	Vision
3. Oculomotor	Motor	Levator, pupil constriction, SR, MR, IR, IO
4. Trochlear	Motor	SO
5. Trigeminal: ophthalmic maxillary mandibular	Both	Lacrimal, frontal, nasociliary
6. Abducens	Motor	LR
7. Facial	Both	Facial expression Facial sensation
8. Vestibulocochlear	Sensory	Hearing, equilibrium
9. Glossopharyngeal	Both	Taste, salivation, motor to pharynx
10. Vagus	Both	Visceral sensory, motor to pharynx, and larynx
11. Accessory	Motor	Larynx, sternocleidomastoid, and trapezius
12. Hypoglossal	Motor	Tongue

Extraocular Muscles (EOMs)

The **annulus of Zinn** lies just inside the orbital apex. This is the site of connection of the muscles to the bony head. The EOMs, except the inferior oblique, originate at the annulus of Zinn and course forward through orbital fascial tissue to attach to the globe. Each muscle turns into a variable amount of tendon as it attaches to the globe in the various locations. The muscles have different types of tissue allowing for quick bursts of activity (saccades) and slow precise movements (pursuit) (von Noorden, 1985). The muscle fibers are in a constant state of readiness. That is, they are always ready to fire. They have **tonus.**

With the head erect, the eye sits in the orbit with its fovea pointed straight ahead at the horizon and assumes the **primary position.** The **visual axis** is the imaginary line between the fovea and the object of regard.

The **muscle plane** includes the center of rotation of the globe, the particular muscle's insertion on the globe, and the muscle's origin. The muscle contracts back toward its firmly fixed origin and pulls the globe around its center of rotation, from its insertion.

Imagine a yoyo on its side as you pull the string backwards. The front of the yoyo (the corneal apex) will rotate around its center towards the finger pulling the string (the origin). If everything is the same—except that the yoyo it tilted halfway up on its side—and you pull back, the front of the yoyo makes a funny combination of movements. It moves back, but less efficiently, and also moves up some. *So even though you are pulling the "same muscle," it's doing a different action.* The muscle action changes when the visual axis and muscle plane *do not* coincide.

This explains why the medial rectus (MR) and lateral rectus (LR) have only one action: their muscle planes naturally coincide with the eye's visual axis see (Figure 1.1). The mus-

EOM Abbreviations:
  Medial Rectus—MR
  Lateral Rectus—LR
  Superior Rectus—SR
  Inferior Rectus—IR
  Superior Oblique—SO
  Inferior Oblique—IO

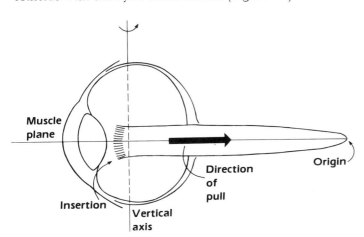

Muscle plane

Insertion

Direction of pull

Origin

Vertical axis

Figure 1.1 Relationship of muscle plane to visual axis of MR.

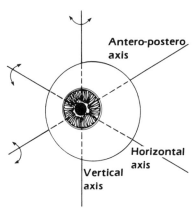

Figure 1.2 Axes of the globe: horizontal, vertical and antero-postero.

cle planes of the other EOMs [superior rectus (SR), inferior rectus (IR), superior oblique (SO), inferior oblique (IO)] do *not* naturally coincide with the visual axis when the eye is in the primary position. When those cyclovertical muscles contract, the movement is a combination of vertical action (around the horizontal axis), horizontal action (around the vertical axis), and the torsional action (around the antero-postero axis) (see Figure 1.2). Remember that the muscles do not "push" the eye around; they contract back toward their fixed origin, pulling the eye with it from its insertion on the globe.

The cyclovertical muscles have three actions. The **primary action** of a muscle is the main action occurring when the eye is in the primary position. The primary action increases when the eye is in ABDuction. The **secondary action** of a muscle has less influence but increases when the eye is in ADDuction. The **tertiary action** is minor, but is either ADDuction or ABDuction.

Medial Rectus (MR)

The MR is the "strongest" EOM for two reasons. First, its insertion is more anterior on the globe than the other EOMs and therefore it has significant **wrap-around effect**. And second, the MR weighs the most; it's the beefiest muscle. Humans use the MR a lot (for reading) which probably accounts for its advanced development.

The **medial rectus** inserts 5.5 mm behind the limbus on the medial side of the globe (see Figure 1.3). This muscle is an average of 41 mm long with 4 mm of tendon. The average width of the tendinous insertion on the globe is 10.5 mm. Its origin is at the back of the orbit at the annulus of Zinn, and the MR is innervated by the inferior division of the oculomotor nerve (CN III). Its function is to rotate the eye medially towards the nose: it ADDucts the eye. With the eye in primary position the MR muscle plane coincides perfectly with the eye's visual axis. The MR is slightly more effective in downgaze, the reading position, and when the eye fixates at near positions.

Lateral Rectus (LR)

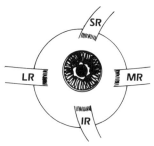

Figure 1.3 Position of insertions of the four rectus muscles, right eye.

A CN VI palsy results in a LR palsy and nothing else.

The **lateral rectus** also works horizontally about the vertical axis. But because it inserts on the globe on the lateral, temporal side, it rotates the eye temporally: it ABDucts the eye (see Figure 1.3). It inserts 7.0 mm behind the limbus, and its average length is 40.5 mm in total length with 8.5 mm of tendon. The average width of its insertion is 9.5 mm, and the muscle originates at the annulus of Zinn. It is innervated by the abducens nerve (CN VI) and is the only EOM innervated by that nerve. The LR is the only rectus muscle whose blood is supplied by one anterior ciliary artery; the other three rectus muscles have two each. The oblique muscles have none. The LR's muscle plane coincides perfectly with the visual axis when the eye is in the primary position and, therefore, has only one action. The LR works most effectively during distance fixation and also in upgaze (in contrast to the MR).

Cyclovertical Muscles

The muscle planes of the remaining four cyclovertical muscles do not coincide with the visual axis and, therefore, each muscle has horizontal, vertical, and torsional capabilities. Although the muscle plane can't change much, the visual axis can. Thus depending on where the eye looks, these actions change. In some positions, the muscle plane and visual axis will be closer to lining up than in other positions.

Superior Rectus (SR)

The **superior rectus** originates at the annulus of Zinn and travels forward above the globe and superior oblique muscle, but underneath the levator lid muscle. It inserts 7.7 mm behind the limbus on the superior side of the globe, farther back than any other rectus muscle (see Figure 1.3). The average length of the SR is 42 mm. The average length of its tendon is 5.5 mm and its tendon width is 10.5 mm. Innervation to the SR is by the superior division of the oculomotor nerve.

When the eye is in the primary position, the SR muscle plane forms an angle of 23° with the eye's visual axis (see Figure 1.4). When the eye ABDucts exactly 23°, the visual axis coincides with the SR muscle plane. At this position, when the SR *contracts,* the eye goes straight up. The function of the SR is pure elevation, with the eye ABDucted 23°. The SR *alone* may be tested when the patient is instructed to elevate the eye from the ABDucted position. Elevation is the **primary** action of the SR and like all cyclovertical actions, the primary action increases in ABDuction and Decreases in ADDuction. When the eye rotates exactly 67° into ADDuction, the visual axis has moved to make a 90° angle with the SR muscle plane. When the SR contracts at this position, it has *no* effect on elevation but rotates the globe about the antero-postero axis. It intorts the eye so that the 12 o'clock position of the cornea rotates nasally. (The right eye's 12 o'clock position moves clockwise when viewed from the front; the left eye's moves counterclockwise.) Intorsion is the **secondary** action of the SR. The insertion of the SR on the globe is anterior to its equator (the imaginary ring that cuts the eye in half front to back), and is relatively temporal to its origin at the annulus of Zinn. When the SR contracts, it pulls the front of the eye nasally and slightly ADDucts the eye. ADDuction is the **tertiary action** of the SR.

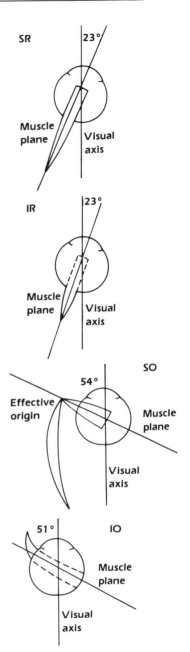

Figure 1.4 Four cyclovertical muscles illustrating angle between muscle plane and the visual axis with the eyes in the primary position.

Primary action increases in ABDuction.

Primary action decreases in ADDuction.

Inferior Rectus (IR)

SR Actions:
1. Elevation
2. Intorsion
3. ADDuction

IR Actions:
1. Depression
2. Extorsion
3. ADDuction

The **inferior rectus** is innervated by the inferior division of the oculomotor nerve (CN III). Like the SR, it originates at the annulus of Zinn and travels downward, then forward, beneath the globe and inserts relatively temporal to its origin, 6.0 mm behind the limbus (see Figure 1.3). Its average length is 40 mm, its tendon 5.0 mm, and its width 10.0 mm at the insertion. The IR works closely with the IO, the lower lid eye muscles, and the Ligament of Lockwood—the hammock-like structure that helps support the globe. For this reason, complications involving the lower lid can arise following "excess" surgery on the IR (Helveston, 1977).

The muscle plane of the IR forms an angle of 23° with the visual axis when the eye is in the primary position (as with the SR) (see Figure 1.4). When the eye ABDucts 23° the IR muscle plane coincides perfectly with the visual axis. If the eye looks down, the depression is caused only by the IR. The **primary action** of the IR is depression. This action increases in ABDuction and Decreases to nonexistence in ADDuction. To see if the IR is working properly, direct the patient to look 23° into ABDuction and then to look down. If the eye looks down from that ABDucted position, the IR is working properly. When the eye ADDucts exactly 67° and the IR contracts, no vertical movement is being generated because the new visual axis position now forms a 90° angle with the IR muscle plane. The only action from ADDuction is movement about the antero-postero axis. When the IR contracts the eye will extort. That is, the 12 o'clock position rotates temporally. (The right eye rotates counter-clockwise when viewed from the front, the left eye rotates clockwise.) Extorsion is the **secondary action** of the IR. Like the SR, the IR insertion on the globe is anterior to the equator and relatively temporal to its origin. When the IR contracts it has a slight ADDucting effect. The **tertiary action** of the IR is ADDuction. So the SR and IR work together as ADDuctors but against each other as torters and vertically acting muscles.

Oblique Muscles

The distinguishing factor of the two oblique muscles is that their force of action is forward because their real origin (IO), or **effective** origin (SO), is anterior to the muscle's insertion on the globe.

The two oblique muscles are mainly responsible for the torsional movements of the globe and for maintaining a forward pulling force on the globe. If the four recti muscles relax as the two oblique muscles contract, the eye would literally pop forward (the Double Whammy syndrome-Geeraets, 1976). Oblique muscles are mainly responsible for maintaining the eye in the upright 12 o'clock position, despite the position of the head. When the head tilts to the right shoulder, both eyes rotate to the left shoulder: the right eye intorts, the left eye extorts. The muscle planes of both oblique muscles do not coincide with the visual axis when the eye is in the primary position. These cyclovertical muscles also have three actions which vary depending on the eye's position.

Superior Oblique (SO)

The **superior oblique** muscle is innervated solely by the trochlear nerve (CN IV) and is extremely susceptible to mild trauma. It is the eye's longest muscle and has the longest tendon. The SO originates at the back of the orbit at the annulus of Zinn, and travels forward along the medial-superior side of the orbit. It passes through the trochlea, it reflects back towards the globe, fans out, and inserts onto the top of the globe slightly behind the equator underneath the SR (see Figure 1.5). Despite its origin at the annulus of Zinn, the SO's *effective* origin is at the trochlea. When the SO contracts, it pulls the top of the globe forward and towards the trochlea. This reflected part of the muscle is mostly tendon. The average length of the entire SO is 59.5 mm, with 19.5 mm of that being tendon. It fans out to a variable width of 7 to 18 mm with the temporal side inserting 13.8 mm behind the limbus and the medial side, 18.8 mm behind the limbus. Mechanical limitations in and around the trochlea can limit the effectiveness of the SO.

The muscle plane of the SO is formed by its reflected portion, including the effective origin at the trochlea. The SO muscle plane forms a 54° angle with the visual axis of the eye when it is in the primary position (see Figure 1.4). When the eye ABDucts exactly 36°, the visual axis forms a 90° angle with the SO muscle plane. When the SO contracts from this position, it intorts the globe around the antero-postero axis. (The right eye rotates clockwise when viewed from the front; and the left eye counter-clockwise.) The **primary action** of the SO is intorsion. With the globe ADDucted exactly 54°, the visual axis and muscle plane coincide precisely. When the SO contracts with the eye in that position, its function is pure *depression*—it pulls the top of the globe forward toward the trochlea. The primary action Decreases in ADDuction. The **secondary action** of the SO is depression (even though it is located superiorly on the globe). The SO inserts slightly behind the equator. With the effective origin relatively medial to its insertion on the globe, the SO pulls the back of the eye nasally towards the trochlea when it contracts. This results in the front of the eye ABDucting about the vertical axis. Thus, the **tertiary action** of the SO is ABDuction.

Inferior Oblique (IO)

The **inferior oblique** muscle is unique in that it is the shortest muscle, with an average 37 mm, and has the

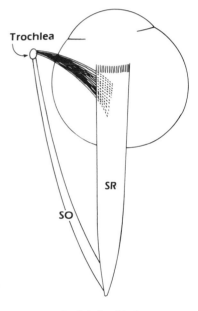

Figure 1.5 Relationship between SO and SR of the right eye.

SO Actions:
1. Intorsion
2. Depression
3. ABDuction

IO Actions:
1. Extorsion
2. Elevation
3. ABDuction

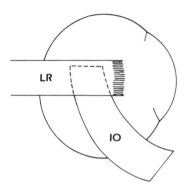

Figure 1.6 Relationship between IO and LR of the right eye.

shortest tendon, with an average of 1 mm. The IO inserts very close to the fovea and is the only EOM whose true origin is NOT at the annulus of Zinn. The IO originates at the posterior lacrimal crest of the infero-nasal orbital rim margin. The muscle travels back in the orbit, underneath the globe and IR (but within Lockwood's ligament), and curves up underneath the LR. Its posterior edge inserts 1 mm anterior to and below the fovea. The anterior edge inserts approximately 17 mm behind the limbus (see Figure 1.6). The IO is innervated by the inferior division of the oculomotor nerve (CN III).

The muscle plane of the IO forms a 51° angle when the visual axis of the eye is in the primary position (see Figure 1.4). When the IO contracts, it pulls its insertion towards its origin in the inferior nasal orbital rim. With the eye ABDucted exactly 39°, the visual axis forms a 90° angle with the IO muscle plane. Contraction of the muscle will then result in pure extorsion of the globe about the antero-postero axis. (The right eye rotates counter-clockwise when viewed from the front; the left eye clockwise.) The **primary action** of the IO is extorsion, which Decreases in ADDuction. With the globe ADDucted exactly 51°, the visual axis coincides exactly with the IO muscle plane. Contraction of the IO in this position would result in pure elevation. The **secondary action** of the IO is elevation even though the IO is located below the globe. The IO inserts behind the equator and relatively temporal to its origin. When it contracts, the IO pulls the back of the globe medially so that the front of the eye ABDucts. Thus, the **tertiary action** of the IO is ABDuction.

Table 1.2 outlines important information about each EOM. Table 1.3 shows the actions of, and where to test each

Table 1.2 Muscle Information

Muscle	Length	Limbus to Insertion	Muscle Plane Angle
MR	41 mm	5.5 mm	0°
LR	40.5 mm	7.0 mm	0°
SR	42 mm	7.7 mm	23°
IR	40 mm	6.5 mm	23°
SO	59.5 mm[a]	13.8/18.8mm	54°
IO	37 mm	17 mm	51°

[a]SO total length, 19.5 mm of it is tendon

Table 1.3 Muscle Actions

Muscle	Actions: 1°/2°/3°	Test
MR	Only ADDucts	ADDuction
LR	Only ABDucts	ABDuction
SR	elevation/intorsion/ADDuction	Up/Out
IR	depression/extorsion/ADDuction	Down/Out
SO	intorsion/depression/ABDuction	Down/In
IO	extorsion/elevation/ABDuction	Up/In

EOM. As previously stated, to test the SR you instruct the patient to ABDuct 23° and then elevate the globe. Although the SR itself is an ADDuctor, its function is tested in the *out* and up position. While the SR and IR can be tested by asking the patient to perform the primary action of each (elevation and depression, respectively), the SO and IO cannot be tested in this way. You cannot ask the patient to intort the eye using the SO! You can, however, ask the patient to show the vertical action of the SO and IO. To do this, you first have the patient take the eye into ADDuction, then elevate (IO function) or depress (SO function).

The mnemonic "SIN RAD" will help you remember the torsional and horizontal actions of the cyclovertical muscles. "SIN" means that the Superior muscles INtort; so the two inferior muscles must extort. "RAD" means that the Recti muscles ADduct; so the obliques must ABDuct.

Descriptive Muscle Terms

The **agonist** muscle is the prime mover for a desired direction of gaze. The **antagonist** muscle of the same eye works directly against the agonist. The MR is the agonist for AD-Duction; the LR, its direct antagonist. A muscle in the same eye that helps another muscle accomplish a particular action is called a **synergist** muscle. So both the SO and IO are synergists for ABDuction with the LR. These same two muscles are antagonists for torsional and vertical action. Table 1.4 shows the synergist-antagonist relationship between the six EOMs for horizontal, vertical, and torsional actions.

Yoke muscles are pairs of muscles (one in each eye) that work together to achieve a desired version movement. **Duction** is movement of one eye; **version** movement of both eyes in one direction. **Dextroversion** is movement of both eyes to the right; **levoversion** is movement of both eyes to the left. *Supra*version for upgaze and *infra*version for downgaze are seldom used. To achieve dextroversion, the yoke muscles used are the right LR (RLR) and left MR (LMR). To achieve fixation up and to the right, the RSR elevates the OD in ABDuction, and the LIO elevates the OS in ADDuction. To

Table 1.4 Synergists/Antagonists

Muscle	Synergists	Antagonists
MR	SR, IR	LR, SO, IO
LR	SO, IO	MR, SR, IR
SR	Elevation: IO	IR, SO
	Intorsion: SO	IR, IO
	ADDuction: MR, IR	LR, SO, IO
IR	Depression: SO	SR, IO
	Extorsion: IO	SR, SO
	ADDuction: MR, SR	LR, SO, IO
SO	Intorsion: SR	IO, IR
	Depression: IR	SR, IO
	ABDuction: LR, IO	MR, SR, IR
IO	Extorsion: IR	SO, SR
	Elevation: SR	IR, SO
	ABDuction: LR, SO	MR, SR, IR

	RSR	LIO			RIO	LSR
	RLR	LMR			RMR	LLR
	RIR	LSO			RSO	LIR

Figure 1.7 Pairs of yoke muscles responsible for moving eyes in to various positions of gaze.

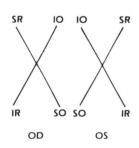

Figure 1.8 Schematic drawing of muscles responsible for moving eyes into tertiary positions.

achieve down and left gaze, the LIR depresses the left eye in ABDuction, and the RSO depresses the OD in ADDuction (see Figure 1.7).

Laws Governing Eye Movements

Two basic EOM laws govern how innervation is supplied to an agonist, its antagonist, and yoke. **Sherrington's law of reciprocal innervation** applies to the agonist and antagonist of *one* eye. Every unit of innervation to the agonist is accompanied by a reciprocal amount of relaxation to the antagonist muscle. As the MR contracts a given amount, its antagonist, the LR, relaxes a reciprocal amount.

Hering's law of simultaneous innervation applies to the yoke muscles of each eye. The fixing eye determines how much innervation goes to the agonist of that eye; an equal amount of innervation then goes to its yoke in the other eye. When all of the muscles are healthy and working at normal levels, the eyes remain parallel, regardless of where they turn together.

When one muscle is weak, or *palsied*, the system is no longer perfectly balanced. For example, if the RMR were palsied, more innervation than usual would be required to move the eye into ADDuction. Its direct antagonist, the RLR, would then require *less* innervation to move the eye into ABDuction because it would be pulling against a weak muscle. The yoke muscles of the RMR and RLR would be affected also because they would receive equal, but inappropriate, innervation. If the palsied eye were fixing and moving into the field of action of the palsied RMR, its yoke (the LLR) would receive the same amount of "extra" innervation and would overshoot into ABDuction. In dextroversion, the RLR's yoke (the LMR) would receive the same amount of innervation and would appear underactive because its own direct antagonist is a healthy LLR. (The so-called primary and secondary deviation and inhibitional palsy of the contralateral antagonist will be described in detail in Chapter 3.) After a period of time, a muscle palsy often recovers to some extent. Conversely, the EOMs may start contracting and changing, causing additional limitations.

Version testing assesses how well a pair of yoke muscles work together. When looking up and to the left, the RIO and LSR should pull both eyes up together to the desired position of gaze. Each eye would have moved with equal speed and smoothness and would be at the same height with respect to each other. This would be a normal version movement. If one eye becomes the fixing eye and the other slows down and never achieves the same position as the fixing eye, an underactive muscle is indicated. Ductions of the eye with the apparent underaction should be performed. If the weakness is still apparent, there is either a restriction holding back the eye, or possibly a severe muscle palsy. If instead the non-fixing eye had sped up and passed the desired position during testing, the muscle is considered overactive. You then must determine if it is a **primary overaction** or a **secondary overaction**. As you will see, overaction and underactions are graded in various ways. (von Noorden, 1985).

Primary versus secondary overactions: A secondary overaction is one that is due to an underactive antagonist. A primary overaction is one that occurs without an antagonist underaction; it simply occurs.

References

Bajandas, F.J., *Neuro-Ophthalmology Board Review Manual.* Thorofare, New Jersey: Slack, Inc, 1980.

Gay A.J., et al., *Eye Movement Disorders.* St. Louis: C.V. Mosby 1974, p 1.

Geeraets, W.J., *Ocular Syndromes.* 3rd ed. Philadelphia: Lea & Febiger, 1976, p 145.

Helveston, E.M., *Atlas of Strabismus Surgery.* 2nd ed. St. Louis: C.V. Mosby 1977, p 64.

von Noorden, G.K., *Burian-von Noorden's Binocular Vision and Ocular Motility.* 3rd ed. St. Louis: C.V. Mosby, 1985.

CHAPTER 2

Physiology of Eye Movements

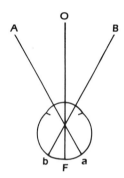

Figure 2.1 Monocular projection: Fovea fixing on the object of regard, point B to right falling on left half of retina (b) and point A to left falling on right half of retina (a).

Figure 2.2 The globe shifting fixation from the object of regard to point B to the right. The front of the eye rotates to the right as the back of the eye (the fovea) rotates to the left in order to pick up fixation (B).

The study of ocular motility and binocular vision is the study of both eyes working together.

Binocular Vision

Monocular projection is either learned or innate (present at birth.) (See Figure 2.1.) The fovea fixates upon the **object of regard** (O). The image of a second object located in space to the *right* of the object of regard (B) falls on the *left* half of the retina (b), or to the left of the fovea. Our brain either innately localizes (thinks) the second real object is actually to the right, or learns quickly of its true location by trial and error. Stimulation of the left half of the retina results in the visualization of an object to the right.

To fixate on the second object located to the right, the back of the eye must move to the left so the image of the object will fall on the fovea. As the fovea moves to the left, the front of the eye rotates to the right to pick up fixation (see Figure 2.2).

If the "amount" of image sweeping across the retina coincides with how much movement the brain thinks the EOMs are causing, NO movement of the environment is perceived. Use your finger to push on your own eye, to move an object across its retina. The brain does not receive any coincident EOM movement information and movement of the environment is perceived!

Binocular projection involves the foveas of both eyes fixing on the same object in space, **bifoveal fixation.** When this occurs in a normal patient, the brain receives a slightly different view from each eye because of the three-inch separation between them. Despite this disparity of images, the views are similar enough to be fused, (blended together) and to be seen as one image in the brain. When horizontally disparate images can be fused because there is only slight disparity, **stereopsis** results—the object is seen in depth. But bifoveal fixation results in more than just stereopsis.

Retinal Correspondence

During bifoveal fixation, the image of an object located in space to the right still falls on the left half of the retina in each eye (see Figure 2.3). The image falls on the *nasal* half of the right eye, and on the *temporal* half of the left eye. The nasal fibers of the two eyes cross at the optic chiasm in such a way that the right eye's nasal fibers and left eye's temporal fibers come together in the left optic tract. These fibers run along the visual pathway back to the occipital cortex, the back part of the brain where vision occurs. It's this crossing of fibers at the chiasm that allows the temporal fibers of one eye to correspond with the nasal fibers of the other. These fibers are stimulated by the image of an object located in the periphery of our vision. The crossing of fibers at the chiasm allows correspondence between them.

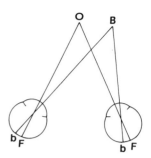

Figure 2.3 Binocular projection and NRC: Bifoveal fixation on the object of regard and an object to the right (B) falls on the left halves of both retinas (b).

At the foveas, vision is so good that there is a "point-to-point" correspondence. In the periphery of the retina, because acuity is poorer, there is more of an "area-to-area" correspondence. Though the large blurry areas in the periphery do not line up perfectly, the brain does not notice this discrepancy because of the poor acuity. When the fovea of one eye corresponds to the fovea of the other, this is a normal phenomenon called **normal retinal correspondence** (NRC).

The **horopter** is an imaginary arc defined as the group of points in space that fall on corresponding retinal points in each eye and therefore seen singly. The horopter "moves" every time a new point is binocularly fixated upon. To you, the horopter would appear as a straight line perpendicular to your combined visual axes. It is actually an arc however, because of the curve of the retinas (see Figure 2.4).

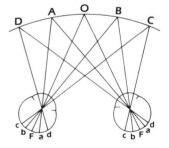

Figure 2.4 Horopter: Bifoveal fixation with a series of points in space that stimulate corresponding retinal points in each eye.

Physiologic Diplopia

Every object on the horopter appears to be the exact same distance away; no depth perception results when objects lie on the horopter. The image from an object that lies significantly in front of the horopter will fall on **bitemporal** retina (see Figure 2.5). The image will fall on the *temporal* half of each retina and therefore *NOT* be on corresponding retinal points. This results in diplopia. Because each eye's monocular projection is always at work, the right eye will see the image of the object nasally (as coming from the left); the left eye will also see the image nasally (but as coming from the right). The right eye is responsible for the left image and the left

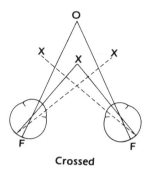

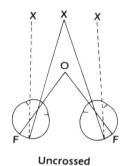

Figure 2.5 *Physiologic diplopia: Distance fixation with crossed diplopia of the near object (top). Near fixation with uncrossed diplopia of the distant object (bottom).*

eye is responsible for the right image. The images have "crossed" over; so bitemporal stimulation results in *crossed diplopia*.

Fixate at a far off door knob, and hold one pencil up directly in front of you. You will see two pencils. Continue fixing on the door knob and close your right eye. The *left* pencil disappears. Then close only the left eye so that the *right* pencil disappears. You are perceiving crossed diplopia of the pencil, and this is perfectly normal.

Similarly, when fixing on one object and a second object is located *behind* the horopter, the image of the second object will fall on what is called **binasal** retina. The right eye will perceive the image as coming from the temporal-*right* side; the left eye will perceive it as coming from the temporal-*left* side. The right eye is responsible for the right image, the left eye for the left image. So the images do *not* cross over; they are "uncrossed". This is **uncrossed diplopia.** (see Figure 2.5). This phenomenon can be observed by fixating on the pencil and noticing the two doorknobs behind, uncrossed. Again, this is perfectly normal.

These two types of **physiologic diplopia** (crossed and uncrossed) are normal phenomena in people with normal binocular vision. Some people become aware of physiologic diplopia and become alarmed by it, so any patient complaining of binocular diplopia should have physiologic diplopia ruled out *first*.

Panum's Fusional Space

There is a small region in front of and in back of the horopter where, although images from it fall on binasal or bitemporal retina, they are mentally fused anyway. The images are disparate, but fusible. When the images fall on slightly bitemporal retina but are fused in this way, it appears to be a single image but *in front of* the horopter. When the images fall on slightly binasal retina from the space just in back of the horopter, fusion occurs and the new single image appears to be *in back of* the horopter. When slightly disparate retinal images (images falling on *non*-corresponding retinal points), fall in the small "fusional space" in front of and in back of the horopter, stereopsis occurs (see Figure 2.6). This space is called **Panum's fusional space**; any object in it will be seen in depth *if it is not* on the horopter. (Remember that images of objects located on the horopter itself fall on corresponding retinal points and therefore are all perceived as being the same distance away.) Slight vertical disparity does NOT result in stereopsis. Panum's fusional space only exists in the horizontal plane.

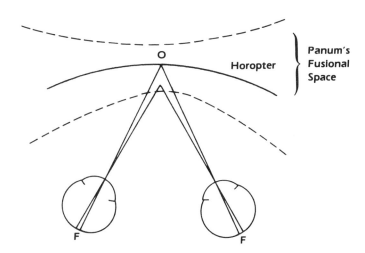

Figure 2.6 Panum's fusional space in the horizontal plane surrounding the horopter. Object points within Panum's fusional space but in front of the horopter seen singly and in depth in **front** of the object of regard (shown). Object points within Panum's fusional space but behind the horopter are seen singly and in depth but **behind** the object of regard.

Confusion and Diplopia

When someone with normal, straight eyes and NRC suddenly develops and eye turn, two phenomena occur. First, the deviated fovea is now pointed at some object other than the object of regard of the fixing eye. The retinal points on each fovea still correspond so the brain momentarily tries to fuse the two different images together. **Confusion** results when trying to superimpose two dissimilar images. Almost immediately, the deviated fovea's image is ignored by the brain in favor of the object of regard of the fixing eye. **Foveal suppression** of the deviated eye begins to occur. Suppression only occurs when both eyes are open and when strabismus is present; suppression only occurs under binocular conditions. If the eye could straighten itself, bifoveal fusion could reoccur and suppression would not be necessary to rid the brain of confusion. (An intermittent tropia does this.)

The second phenomenon that occurs when strabismus affects a normal person is **pathologic diplopia**. This happens when the image of a single object of regard falls on noncorresponding retinal points. When a RET occurs, the left eye continues to fixate with its fovea (see Figure 2.7). The right eye is crossed in such a way that the object of regard falls on a nasal retinal point of the right eye (N). When an image falls on nasal retina, it is perceived as having come from the temporal side—the right side of the right eye. So uncrossed diplopia results from a RET.

When a LXT occurs (see Figure 2.8), the right eye continues to fixate with its fovea. The left eye turns outward so that the object of regard now falls on a temporal retinal point of the left eye (T). The object is perceived by that deviated eye as having come from the nasal (right) side of the left eye. So crossed diplopia results from a LXT. A young patient can

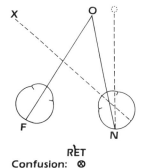

RET
Confusion: ⊗
Diplopia: O ○

Figure 2.7 Pathologic diplopia: Uncrossed diplopia resulting from a RET where the object of regard (O) falls on nasal retina (N) of the deviated right eye and is seen temporally to its actual location. The left eye still sees the object of regard (O) straight ahead so that two images are seen.

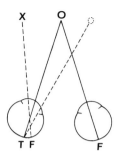

LXT
Confusion: ⊗
Diplopia: O ◌

Figure 2.8 Pathologic diplopia: Crossed diplopia resulting from a LXT where the object of regard (O) falls on temporal retina (T) of the deviated left eye and is seen nasally to its actual location. The right eye still sees the object of regard (O) straight ahead so that two images are seen.

Worth's Three Grades of Fusion:
1. Superimpositioning
2. Motor Fusion
3. Stereopsis

learn to ignore the extra image so that only one image is seen. Suppression of the image occurs only under binocular conditions (see Chapter 4).

Fusion and Fusional Movements

Although humans have two eyes and therefore see two images, those images are similar and can be blended mentally in the brain to be seen as one good image (**sensory fusion**). Worth describes three grades of fusion to explain this phenomenon.

Grade 1 fusion is **superimpositioning** of dissimilar objects. When the corresponding foveas are stimulated, even if by different objects, sensory fusion should take place. Superimpositioning can be tested on a **haploscopic instrument**, a special device that shows one picture to the right eye and a completely different picture to the left eye. For instance, one fovea is shown a lion, the other a cage, and the patient is asked to superimpose the lion in the cage (see Figure 2.9). This would be grade 1 fusion. Despite normal fusion, it is difficult to fuse dissimilar images. Even though a lion "belongs" in a cage, rapid, alternate central suppression often causes the central bars of the cage to disappear or the lion's midsection to fade out.

Grade 2 fusion, according to Worth, is **motor fusion**. Motor fusion is the fine-tuned movements that are done in the interest of maintaining sensory fusion. It is tested by assessing vergence amplitudes. This can be done with either a haploscopic instrument, but most frequently, it is done by measuring vergence amplitudes under normal seeing conditions with a prism bar or Risley prism. When using a haplo-

Figure 2.9 Grade 1 fusion: Image that brain would perceive when superimpositioning was present and the lion was presented to one fovea and the cage was presented to the other fovea on a haploscopic instrument.

Figure 2.10 Grade 2 targets: Butterfly is monocular check point on left slide. Cat's tail is monocular check point on right side.

scopic instrument, similar targets are used but with check points on each picture (see Figure 2.10). The check points help to determine when one eye's image disappears from view (when fusion no longer occurs). If only one cat is seen but that cat has no tail, one eye is being suppressed. The actual measurements of motor fusion are described in Chapter 6.

Worth's grade 3 fusion is **stereopsis.** Similar pictures that are slightly disparate but fusible are used on a haploscopic instrument to test stereopsis. Obviously, grade 2 fusion (good vergence amplitudes) is *not* a prerequisite for grade 3 fusion. Normal stereo (grade 3) could exist without any motor fusion (grade 2). Superimpositioning (grade 1) is difficult for any patient, and some are barely able to appreciate this despite having perfect motor fusion (grade 2) and/or stereopsis (grade 3). While Worth's grades of fusion are helpful in understanding the sensory/motor fusion mechanism, the graded numbers can be misleading.

Fusion Potential

It is often necessary to determine if a patient *fuses,* or if there is **fusion potential.** The fusion being referred to is **sensory fusion.** Good fusion potential implies that if the eyes become straight—either by surgery, refractive correction, or prisms—the patient would then have good sensory fusion. To measure sensory fusion in a patient with a constant eye turn, manipulate the strabismus and seeing conditions so that the eyes are in the position where they are *most likely* to fuse. Then measure the sensory fusion.

Measuring sensory fusion requires a subjective response from the patient. The patient has to tell you which circle is closest, how many lights there are, or if the lion is in the cage. Try these questions on a two-year-old. The only way to test for SENSORY fusion in a young child or in a patient whose subjective responses cannot be relied upon is by testing their MOTOR fusion. Remember that motor fusion

Good MOTOR fusion is an indication of good SENSORY fusion.

is done in an attempt to maintain sensory fusion; so if motor fusion is exerted, it must indicate that sensory fusion exists. While sensory fusion often can't be measured directly, it can be measured indirectly by measuring motor fusion. The methods and tests for measuring sensory and motor fusion are described in chapters 4 and 6.

Convergence and Divergence

Five Types of Convergence:

1. Tonic
2. Proximal
3. Accommodative
4. Fusional
5. Voluntary

The lateral walls of the orbits form a 90° angle, so the eyes naturally diverge. However, several different types of convergence bring the eyes from the divergent position caused by the orbit's structure into the straight-ahead, fusible position.

Tonic convergence is a result of the tonus of the EOMs and exists when the patient is conscious enough to have electrical activity firing in the muscles. Tonic convergence can exist under monocular conditions; it does not require binocularity.

Proximal convergence is the result of a person's awareness of near. Like monocular proximal accommodation, which exhibits itself while viewing through ophthalmic instruments, proximal convergence can also exist under monocular conditions. Proximal convergence and accommodation are also referred to as "instrument" convergence/accommodation. This is because the artificial viewing distance created by the instrument being used induces the patient to inappropriately converge/accommodate. The instruments (haploscopic devices used for measuring and treating strabismus and binocular vision problems, and instruments such as the ophthalmoscope) are made to optically simulate distance (infinity) fixation and, therefore, do not require any accommodation or convergence. The viewer, however, knows that the object of fixation is only a few inches away and automatically tends to overconverge/accommodate and the result is **proximal** convergence/accommodation.

Proximal convergence is measured by comparing the position of the eyes when measured on a haploscopic instrument set for distance fixation, to the position of the eyes when measured while fixing at distance (20 feet). A patient who measures 3 X by prism and cover in the twenty foot lane and 2 E on the haploscopic instrument, is "overconverging" 5 P.D. (prism diopters) on the haploscopic instrument so the patient is using 5 P.D. (prism diopters) of proximal convergence. Proximal accommodation is estimated by how much minus power in diopters an emmetrope adds to an instrument in order to see the image clearly.

Accommodative convergence is the amount of convergence exerted for each unit of accommodation exerted. To see clearly at near distances, a certain amount of accommodation (measured in diopters) is required. At the same time, the eyes need to converge a certain amount to bifoveally fixate on the near object. The amount of convergence necessary depends upon fixation distance and the person's interpupillary distance, and is measured in P.D. Accommodative convergence can exist under monocular conditions, but provides only a rough approximation for the position of the eyes in preparation for bifoveal fixation.

These three types of convergence—tonic, proximal and accommodative—usually position the eyes close to the straight-ahead position under monocular conditions. If the eyes are *not* straight when both eyes are open, however, diplopia results.

Fusional convergence (or divergence) is necessary to fine-tune that resultant eye position so that bifoveal fixation and fusion can be assumed. Fusional convergence is used to straighten the slightly divergent eyes. The eyes would be *exophoric* or, only turn outward—when fusion is disrupted, as with a cover test. **Fusional divergence** is used to straighten slightly convergent eyes. The eyes would be *esophoric*—or the eye under the cover would be turned inward—when fusion is disrupted. The eyes are only straight when binocular fusion is allowed.

The eyes accommodate the appropriate amount in order to see clearly for a given fixation distance. Fusional convergence/divergence must be done without changing this accommodation. This is termed **relative** convergence/divergence. Progressively larger base-in prisms are placed in front of one eye to measure divergence until the prism cannot be overcome (diplopia results). Progressively larger base-out prisms are used to measure *fusional convergence* until the patient reports blurring. Blurring signifies that the patient's fusional convergence has been exhausted and other types of convergence are being used. The blurring is caused by *over*accommodation in an attempt to use accommodative *convergence*. Patients converge and overcome the base-out prism, but are using accommodative convergence instead of fusional convergence. The target may be single, but it doesn't appear clear. When fusional convergence is deficient and accommodative convergence is used, there is blurring while reading: the classic symptom of convergence insufficiency.

The fifth type of convergence is really a combination of the others. **Voluntary convergence** is done without visual stimuli; it is done by patient choice. This type is usually accompanied by pupillary miosis, so accommodative convergence is certainly exerted (Jampolsky, 1970). Voluntary convergence is an easily learned trick: blind people can do it, and large XTs can do it. Children wanting to torment their mothers and challenge the old wive's tale that crossing your eyes will get them stuck can certainly learn to do it.

Which kind of convergence is measured when the NPC (near point of convergence) is calculated? Tonic convergence is always present. Proximal and accommodative convergence are used because the target is close to the eyes. Voluntary convergence may also be used because the patients are instructed to watch the pencil as it comes closer to their nose. (A blind person with voluntary convergence could follow those instructions.) Fusional convergence may be used, but to an unknown extent. The fact is, when a satisfac-

Fusional vergences (convergence and divergence) are measured by determining how much the eyes can be forced to converge or diverge without changing accommodation.

Cover-uncover test: Each eye is covered, and then uncovered, one at a time. As one eye is covered, the **opposite** eye is watched for movement. As the eye is uncovered, **that** eye is watched for movement as the cover is removed.

The cover forces the opposite eye to be the fixing eye. If that opposite eye already was fixing, it doesn't have to move as the cover is introduced. If that opposite eye was deviated (tropic), it will have to move in order to pick up fixation. So the "cover" part of the test determines if a tropia exists.

Remember that a phoria is an eye turn that exists only when fusion is disrupted such as when one eye is occluded. A patient with a phoria has straight eyes when both eyes are open. The eye only deviates when fusion is disrupted. So **as** the eye is occluded, the eye underneath the cover deviates because binocular fusion is disrupted. When the cover is removed, watch the eye being uncovered for movement. So the "uncover" part of the test determines if a phoria exists.

A **phoria** exists if: A. No movement occurred in the opposite eye as the eye was covered, and B. The eye being uncovered **moved** in, out, up, or down to pick up fixation. A **tropia** exists if: Movement occurred in the opposite eye as the eye was occluded.

An **intermittent tropia** may act as a phoria initially, and then decompensate into its tropic state. When an intermittent tropia tries to regain binocularity when the cover is removed, it takes longer than a phoria does to regain fusion.

NPC (near point of convergence) measurements **DO NOT** assess a patient's ability to comfortably converge. A patient with a good NPC may **still** have convergence insufficiency!

Constant eye turns, **tropias**, are designated as either crossing in, **esotropia, (ET)**, turning out, **exotropia, (XT)**, or turning up, **hypertropia, (HT)**. At **intermittent tropia** is designated by parenthesis around the "T"; intermittent XT is **X(T)**. **Phorias** are designated without a "T" altogether; an esophoria is designated by **E**. Any **near deviation** is designated by a prime after it. A RHT at near, would be **RHT'**.

tory NPC has been measured, you don't know what has been measured.

To determine if a patient's asthenopic symptoms are due to a convergence problem, *fusional* convergence must be tested. An NPC measurement cannot provide this information. A poor NPC tells you that there is probably something wrong; a good NPC tells you nothing. Likewise, *teaching* patients how to improve their NPC—the infamous pencil push-up—just teaches them the trick. It does *not* increase their fusional convergence and therefore does not relieve their symptoms. It *does* give them the idea that "convergence exercises" don't work and it confuses the next person who examines their spectacular NPC, despite symptoms of convergence insufficiency. My best advice here is to use an NPC and its information cautiously, because it doesn't tell you much.

Phorias, Tropias, Intermittent Tropias

Eye turns, or **strabismus** results when the eyes are not working together properly. Eye turns can basically be described in one of two ways: either by the position of the deviating eye (eso, exo, hyper, hypo), or by its degree of control (phoria, tropia, intermittent tropia).

A **phoria** is an eye turn that is in full control and becomes apparent only when fusion is disrupted. Covering one eye interrupts fusion by dissociating the eyes. The cover- uncover test therefore reveals a phoria. Covering one eye momentarily interrupts fusion and allows the deviation to become manifest underneath the cover; uncovering the eye allows fusional vergence to return the eye back to the ortho position.

A **tropia** is a constant eye turn without any ability to control it with fusion. Patients with a tropia have either constant diplopia, or suppression if the tropia was developed early in life. Patients with a tropia and suppression are not aware that they have an eye turn except when looking into a mirror, or when they are directly told that they do. In contrast, patients with a phoria have a built-in mechanism that signals when the eye is beginning to deviate; so the eye never actually *does* deviate. Those with a phoria would be aware of diplopia as soon as the eye *started* to turn. Because of their fusional amplitude, the impending eye turn is overcome and the deviation is controlled. A tropia usually does NOT have the diplopia warning system and never has vergence amplitude to control it if the deviated eye is suppressed.

An **intermittent tropia** sometimes behaves as a phoria, and sometimes as a tropia. The eyes may be straight and fusing one moment; then, without any known provocation, deviate with suppression—and maybe even ARC—the next moment. Because the eye is suppressed when tropic, it is presumed that their straight eyes are the result of fusion. A blink, a change of eye position, intense brightness, or fatigue may provoke that person into going tropic.

Control of the deviation differentiates a phoria from a tropia and from an intermittent tropia. A phoria is always controlled, a tropia is never controlled and an intermittent tropia is only sometimes controlled. Diplopia "tells" a patient that the deviation is becoming manifest and that control is needed. A tropic patient who suppresses has no symptoms or clues to know when to pull his or her eye straight. Diplopia usually causes a **reflexive** vergence movement that is done automatically and without requiring thought.

Treatment is sometimes aimed at making a patient's intermittent tropia behave like a phoria. This might be accomplished by teaching the patient to appreciate diplopia when tropic and appropriate, and by teaching better vergence amplitudes, either convergence or divergence.

Dissociating Factors

Dissociating factors are things that are likely to bring out an eye turn, to break it down into its tropic form. The most extreme dissociating factor is prolonged occlusion. The cover-uncover test provides only a short period of occlusion. Occasionally, this short period is not dissociating enough to break down a phoria. Cross-covering is only a slightly longer form of occlusion, and may still not be dissociating enough to break down some exophorias. Prolonged occlusion, up to 24 hours, is required to dissociate the eyes in these cases. Measurements must be taken without allowing any binocularity, so the patient is instructed to come back to the office wearing the patch. The examiner removes the patch only when ready to measure the size of the deviation and only when the other eye is occluded (see Figure 2.11).

Dissociating Factors:
1. Occlusion
2. Shape differences
3. Color differences

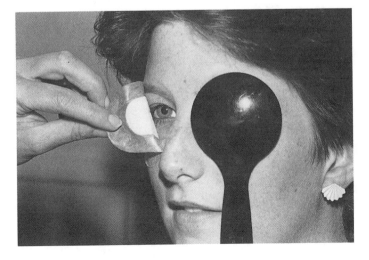

Figure 2.11 After 24-hour occlusion of right eye, the left eye is covered with the occluder **before** the patch is removed so that binocular fusion is not permitted.

Any instrument that makes each eye see a different-colored or different-shaped image is a dissociating factor. The Maddox rod is dissociating because one eye sees a red line while the other sees a white light (different shape and color). The Worth four dot is dissociating because it makes one eye see red and the other see green (different colors). When tests are done in a darkened room, not only is the contrast greater between background and fixation object, there also aren't any peripheral fusion clues. Both the Maddox rod and the Worth four dot test are performed in the dark, but red filters may be used either in the light or in the dark (different color). Graded filters from pale pink (1) to dark red (15) are used to "grade" how easily patients can be dissociated. If the tropia breaks down with only a pale pink filter, fusion is weak. If a very dark red filter is required to break the tropia down, fusion is stronger.

Bagolini lenses for testing retinal correspondence are only slightly dissociating and are used under normal seeing and light conditions. The patient fixates on a bright light and each eye sees a streak of light that is oriented differently (different shape). Merely looking at a light can be dissociating because of the brightness. The stereotest dissociates very little because the images seen are only slightly disparate (different shape). Remember, it is this disparity that is necessary for stereopsis to be appreciated. While stereo-testing is considered the *least* dissociating of all tests, it is sometimes the ONLY test where patients—especially those with exodeviations—will pull their eyes straight and become able to report fusion.

Figure 2.12 Accommodative mechanism: Contraction of the ciliary muscle results in relaxation of the zonules resulting in bulging of the lens (increased convexity). This results in a more plus (+) lens with a shorter focal length.

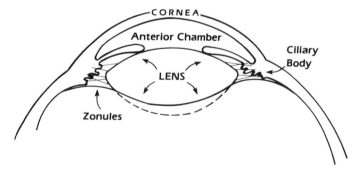

Accommodative Mechanism

The young human eye is equipped with the ability to focus. This ability is measured in diopters and depends upon age and the amount, or amplitude, of accommodation. The young person's lens is pliable, and if it's allowed to go into its natural

natural shape, it will bulge and becomes a very convex, plus lens. The lens is suspended in the eye by the zonules of Zinn, which are connected to the valleys between the ciliary processes of the ciliary body (see Figure 2.12). When the ciliary body *contracts* the zonules can *relax*, the lens can bulge, and accommodation occurs.

References

Jampolsky, A., Ocular Divergence Mechanisms. Transactions of the American Ophthalmology Society 68:730, 1970.

CHAPTER 3

Classifications of Strabismus

Eye turns are classified in a variety of ways. They may be classified by control (phoria, intermittent tropia, constant tropia), time of onset (infantile/congenital, acquired), etiology (traumatic, accommodative), direction of eye turn (eso, exo, hyper, hypo), gaze variation (comitant, incomitant), restrictions, muscle palsies (CN III, CN IV, CN VI), or by the company the eye turn keeps (syndromes).

Comitant vs. Incomitant Deviations

Detecting Incomitant Deviations

The common unit for measuring strabismus is the prism diopter. It is abbreviated **P.D.** or is designated by the symbol: △.

Comitant deviations are eye turns that measure the same amount in all directions of gaze. **Incomitant** deviations are eye turns that *vary* in amount depending on direction of gaze. A patient with a 20 P.D. (prism diopter) exo*phoria* in downgaze and a 20 P.D. exo*tropia* in upgaze is still considered to have a comitant deviation because the *size or amount* did not vary; only the control varied. An incomitant deviation is usually caused by an EOM imbalance, a neurological imbalance, or a restrictive strabismus. Incomitant deviations due to underactive muscles are frequently caused by muscle palsies. These deviations must be carefully documented and their etiologies investigated (see Figure 3.1).

To test for incomitance, measurements must be taken with the patient fixing on a distance accommodative target

		20△RET		
Right gaze	40△RET	20△RET	5△E	**Left gaze**
		20△RET		

Figure 3.1 Incomitant deviation showing largest deviation in right gaze, smallest deviation in left gaze. RLR palsy.

with their correction on. These measurements are taken with the patient fixing in the primary straight-ahead position, the secondary positions (up, down, right, and left), and the tertiary positions (up and right, up and left, down and right, down and left). (See Figure 3.2.) The presence of a primary and secondary deviation should be documented. Prism and cover measurements (see Chapter 6) may be taken on any patient who will hold fixation. Maddox rod measurements (see Chapter 6) may be taken on any patient who has NRC, little or no suppression, and honesty. Distance Krimsky *estimations* are possible when a patient has none of the above. The measurements are considered to be more accurate when taken at distance than when taken at near (Bredemeyer and Bullock, 1968). This is because no additional accommodation occurs that might affect the eye turn; patients wear their glasses to control accommodation. On a baby who will not fixate well in the distance, near Krimsky or Hirschberg measurements must be resorted to (see Chapter 6).

3°	2°	3°
2°	1°	2°
3°	2°	3°

Figure 3.2 Three types of position of gaze. 1° (primary)—straight ahead. 2° (secondary)—up, down, right, and left. 3° (tertiary)—up and right, down and right, up and left, and down and left.

Types of Incomitant Strabismus

A and V patterns are a type of horizontal strabismus that varies in up and down gaze. A patterns are most convergent in upgaze and most divergent in downgaze. V patterns are most convergent in downgaze and most divergent in upgaze. Many normal people have and A or V *tendency*. A minimum of 10 P.D. difference between up and down gaze is necessary to substantiate an A pattern, from an A tendency. A minimum of 15 P.D. is necessary to substantiate a V pattern.

Examples of A and V patterns in eso and exo patients are shown in Table 3.1. V pattern ETs are the most common, followed by V pattern XTs, A pattern ETs, and then A pattern XTs.

To determine the presence of an A or V pattern, measurements are compared between the primary position — 25–30° in upgaze, and 25–30° in downgaze. These measurements are best taken with the head either being moved up for downgaze, or down for upgaze. The 25–30° rotations move the eyes into extreme gaze. Here, the over-and underactions of the muscles will disturb the horizontal eye position significantly enough for documentation.

When testing versions in the presence of an A or V pattern, MR, LR, SR, IR, SO, and IO function should be looked at carefully, and these muscles should be directly treated if they are dysfunctioning. Treatment of A or V patterns caused by EOM over- or underactions should be aimed at strengthening underactive muscles (**resection**) or at weakening overactive muscles (**recession**).

There are three theories about the causes of A and V patterns. The **oblique theory** is the most popular and, indeed,

Any age child who is suspected of having strabismus, should be evaluated immediately by an ophthalmologist.

Table 3.1 Examples of A/V Patterns

	V ET	V XT
Upgaze	10 E	35 XT
Primary	20 ET	20 X(T)
Downgaze	30 ET	10 X

	A ET	A XT
Upgaze	12 ET	ortho
Primary	8 ET	ortho
Downgaze	ortho	15 XT

Three theoretical causes of A/V patterns:
1. horizontal rectus muscle over/under actions
2. vertical rectus muscle over/under actions
3. oblique muscle over/under actions

the oblique muscles are most frequently found at fault in A and V patterns. A few authors have found lid fissure slants to correlate with A and V patterns; bony facial structure anomalies definitely coexist with horizontal incomitance in up and down gaze. (Patients with Crouzon and Apert syndromes have been found to have V patterns.)

The **horizontal theory** maintains that the MR converges the eyes most efficiently in down gaze, and that the LR diverges the eyes most efficiently in upgaze. A patient with a recent LR weakness will therefore have an A tendency. When a basic ET or XT exists, the main deviation and overactive MR or LR will cause a V pattern; and underactive MR or LR will cause A patterns. Treatment of A and V patterns is aimed at the MR and LR only if these muscles are found at fault.

The **vertical recti theory** recognizes that the SR and IR are ADDuctors. If the SR is overactive and only works in upgaze, it would ADDuct the eye in that position: an A pattern. If the SR is underactive, it can't ADDuct properly in upgaze and the eyes will be in a divergent position in upgaze: a V pattern. An overactive IR would ADDuct the eyes too much in downgaze: a V pattern. An underactive IR would cause the eyes to be in a divergent position in downgaze: an A pattern.

The **oblique theory** is regarded as the most likely explanation of A and V patterns. The oblique's tertiary action is ABDuction (although the muscle itself itself is tested in ADDuction, where its vertical action is maximum). An overactive IO causes the eyes to be over-ABDucted in upgaze: a V pattern. An underactive IO causes the eyes to be more convergent in upgaze: an A pattern. Overactive SO over-ABDuct the eyes in downgaze and cause the eyes to be divergent in downgaze: an A pattern. Underactive SO (fairly common because of the frequency of isolated CN IV palsy) cause the eyes to converge in downgaze: a V pattern (see Table 3.2).

Muscle palsies also cause incomitant deviations. When one or more of the 12 EOMs are underactive, the entire system goes out of balance. Each EOM works most efficiently in a particular field of gaze, and a new muscle palsy affects that field of gaze the *most* (causes the largest deviation). The palsied muscle works "least" in the exact opposite

Table 3.2 Causes of A/V Patterns

A patterns may be caused by:
   Overactive: SO, SR
   Underactive: IO, IR, LR, MR

V patterns may be caused by:
   Overactive: IO, IR, MR, LR
   Underactive: SO, SR

Table 3.3 Gazes Affected by EOM Palsy

Muscle	Most	Least
MR	ADDuction	ABDuction
LR	ABDuction	ADDuction
SO	Down/In	Up/Out
IO	Up/In	Down/Out
SR	Up/Out	Down/In
IR	Down/Out	Up/In

field of gaze and, therefore, affects the deviation the *least* (see Table 3.3).

In time (weeks to years), the deviation looses its dramatic incomitancy; there is a **spread of comitance.** This occurs as the other healthy muscles change in response to the palsied muscle and/or as muscle contractures occur. Usually, the **direct antagonist** of the weak muscle becomes overactive, as it is now pulling against a weak muscle. This causes the deviation to spread to the field of the direct antagonist. The yoke of the direct antagonist then receives *less* innervation because the direct antagonist requires less innervation to achieve its desired location. This causes the yoke of the direct antagonist to *appear* weak. With the paretic eye fixating, the yoke of the weak muscle receives *more* innervation because the paretic muscle requires more innervation to move the eye into the desired position of gaze. The spread of comitance spreads *away* from the weak muscle's field of action and toward the opposite field of action (see Figure 3.3).

Muscle contractures also contribute to the spread of comitance when a muscle stops moving freely because of a muscle palsy. The muscles themselves tighten up and, eventually, may not allow the eye to rotate even when pulled with forceps (forced ductions).

The other hallmark of a new muscle palsy is the presence of a *primary and secondary deviation.* The deviation varies, depending upon which eye is fixing. The amount of innervation is determined by which eye is fixing; more innervation is required to get the paretic eye to the desired fixing position. That additional innervation also goes to the "good" eye in accordance with Hering's Law. Those healthy muscles then "overact" causing the (secondary) deviation to be larger in comparison with the deviation measured with the "good" eye fixing.

Duane's retraction syndrome of which there are three types, causes a horizontal deviation that varies in primary, right, and left gaze (see esodeviations and exodeviations). This is thought to be caused by a misfiring, or by a true cofiring of the MR and LR nerves. The patient is often ortho near, or at the primary position, and may fuse with equal vision. Duane's syndrome type 1 causes an increasingly larger ET to the ABDucting side. Type 2 causes an increas-

Figure 3.3 Example of spread of comitance of a LSO palsy. Largest deviation is initially in the down and right position where the action of the LSO would be greatest. The deviation then usually spreads up the right field of gaze toward the action of the LIO (the direct antagonist) (a), across the lower fields of gaze toward the down and left position of gaze (b), and diagonally up and across affecting the up and left field of gaze least and last (c).

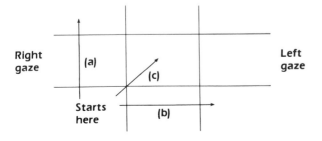

All three types of Duane's retraction syndrome have narrowing of lid fissure and globe retraction during attempted ADDuction.

Type 1: Decreased ABDuction
Type 2: Decreased ADDuction
Type 3: Both (Decreased ABDuction and ADDuction)

ingly larger XT to the ADDucting side. Type 3 is decreased ABDuction and ADDuction of the eye so that the ET increases to the ABDucting side, and the XT increases to the ADDucting side. All three types result in narrowing of the lid fissure and in globe retraction when ADDuction is attempted. Duane's syndrome is predominantly found in females and, most commonly affects the left eye.

Primary overactions of EOMs are NOT secondary to a muscle palsy. They are often responsible for A and V patterns and cause an incomitant deviation in the field of action where the muscle works. Prism and cover measurements in the nine positions of gaze would show a variation in the *hyper*deviation as the field of the overactive muscle is approached.

Pseudo ET due to epicanthal folds is NOT a true form of incomitant strabismus. Epicanthal folds, the fold of skin over the nasal canthus, make a young child *appear* as though he has an ET. This ET appears to get larger when the child looks to right or to left gaze. This is because the ADDucting eye is hidden more by the skin over the medial canthus (see Figure 3.4). Two factors make pseudo ETs due to epicanthal folds very deceiving. First, the eye turn *appears* to worsen in right and in left gaze, which is exactly what actually happens in bilateral LR palsy (a deadly finding in an infant). Second, a child occasionally has asymmetrical epicanthal folds, or doesn't outgrow them with age. While a pseudo ET is completely benign except for temporary appearance, a real ET that *truly* gets larger in right or in left gaze may be due to a potentially life-threatening unilateral or bilateral CN VI palsy. My best advice here is that an ophthalmologist investigate *all* suspected ETs in infants particularly if the ET is described as being incomitant and getting larger when the child looks to the right or to the left.

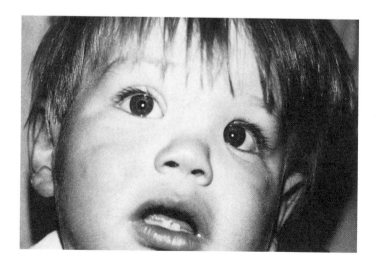

Figure 3.4 *Pseudo ET: epicanthal folds makes child* **appear** *to have an ET. Hirschberg testing indicates no tropia as the light reflexes are both equally centered. Cover-uncover testing confirmed no tropia.*

Restrictive Strabismus

Eye turns may also result when both eyes are not moving freely. When one eye is fixed in place while the other freely rotates, the angle between the eyes will vary and the deviation will be incomitant. Restrictive strabismus is always incomitant.

Restrictive strabismus is any eye turn due totally or in part to one or both eyes being prohibited from freely moving. An individual muscle may be mechanically tethered down, or one or more muscles may be affected by a systemic disease that causes them to become less elastic, less *movable*. Surgical treatment of restrictive strabismus is unpredictable and often disappointing because the muscles are no longer viable and healthy.

Identifying Restrictive Strabismus

Restrictive eye movements are suspected when the eyes do not move freely for version testing, or when an "apparent" muscle underaction exists. If the eye does not move freely when monocular duction testing is done, an ophthalmologist may perform **forced ductions.** The eye is anesthetised, and the globe is bluntly forced to move with a cotton-tipped applicator or grasped with forceps. An eye that does not elevate easily would be suspected of having restrictions of the *inferior* muscles. An attempt would be made to force the eye into upgaze with the forceps. If the eye could not be forced up, or if the tester felt resistance to the movement, the test would be considered a *positive forced duction test*. If the eye moved freely, the patient would have *negative forced ductions.*

Forced duction testing helps determine if the eye is restricted. "Positive forced ductions" indicate restriction. "Negative forced ductions" indicate NO restriction.

Eye deviations that are caused by restrictive strabismus may not be successfully measured by prism and cover measurements, or any other measurements requiring primary gaze head positioning and that each eye pick up fixation alternately. Krimsky measurements may be resorted to in these cases.

Types of Restrictive Strabismus

Grave's ophthalmopathy, (thyroid eye disease) may manifest itself in many nonstrabismic ways. The optic nerve may be compromised when the EOMs become congested. Lid retraction (stare) due to sympathetic over-stimulation of Mueller's lid muscle may require cosmetic surgery. Exophthalmos caused by increased volume of the orbital contents (EOM infiltration) may cause severe corneal exposure keratitis that requires orbital decompression surgery.

Restrictive types of strabismus:
1. Grave's ophthalmopathy
2. Brown's syndrome
3. blowout fracture
4. generalized fibrosis syndrome
5. muscle contractures

Adult patients who require prismatic correction usually prefer a single prism (not split) over their nondominant eye, and base **UP** if possible. A base-up prism permits patients to read in down-gaze through the thinnest part of the prism (the apex), where there will be less distortion.

Restrictive strabismus is caused by tightening and progressive inelasticity of the EOMs. Grave's is a system disease, so *both* eyes will be affected, but often asymmetrically. There may occasionally be an actual CN palsy with or without the restrictions.

The IR and MR muscles are most frequently affected, and are usually the first muscles affected. Classically, this causes a limitation of *upgaze* and *ABDuction*. However, all EOMs may be affected.

Because thyroid eye disease usually affects adults (young females in particular), patients experience diplopia from the strabismus. They may experience a decrease in their field of view because they can no longer freely rotate their eyes. They also exhibit decreased ductions and positive forced ductions. The smaller hyperdeviations often require prismatic correction; because of the variable nature of the disease, temporary press-on Fresnel prisms are useful.

Measurements should be taken in all positions of gaze, if possible (Hess or Lancaster Red-Green is helpful). Fusion when the eyes are in the orthophoric position should be documented regardless of the field of gaze. A diplopia field helps document the exact position and extent of a patient's single binocular vision and diplopia. This is particularly useful pre- and postoperatively. Thyroid patients often have convergence insufficiency (Moebius' sign) and should be treated appropriately.

Brown's syndrome may mimic IO palsy. Forced duction testing would be **positive** for Brown's syndrome and **negative** for IO palsy.

Brown's syndrome (Brown's superior oblique tendon sheath syndrome) is a mechanical restriction preventing free passage of the SO through the trochlea. The SO is unable to pass freely through the pulley when the eye looks up and in. This then *appears* to be a decreased elevation in ADDuction by duction testing (the IO appears weak). The patient may have a chin-up head position to put the eyes into downgaze. Although Brown's syndrome is usually congenital (and monocular), it may be acquired through damage—either traumatic or iatrogenic—to the trochlea itself or to the portion of the SO that glides through the trochlea. Brown's syndrome is diagnosed with a positive forced duction test.

Some children with juvenile rheumatoid arthritis have a SO "click" syndrome. When ductions or forced ductions are performed, the eye, at first, will not elevate in ADDuction. But with continued attempts, the SO literally clicks through the obstruction and the eye pops up into position. For this reason, any child with a Brown's syndrome should be evaluated by a pediatrician for juvenile rheumatoid arthritis (Killian, McClain, and Lawless, 1977).

Brown's syndrome does not usually require treatment (except for amblyopia) if the child is fusing in or near primary position or if the condition is cosmetically acceptable.

In Brown's syndrome, the eyes are often straight in the primary position, so fusion is often present and amblyopia is an infrequent finding. A child who looks up often into the IO's field on the affected side will suppress the extra image. The child who does NOT look into that field often will *never* learn to suppress and will be diplopic in that field of gaze as an adult.

Blowout fracture is the result of blunt trauma to the globe and orbit. The orbital rim is strong and absorbs much of the shock of the blow without cracking. The force of the blow is usually directed back toward the orbit's apex and results in the orbital walls fracturing under the pressure. (The orbital floor is most frequently affected, followed by the medial wall.) As the force hits the orbit and globe, the floor cracks. At the same time, the orbital contents are pushed back into the orbit with no where else to go except *into* the crack. Moments later, the pressure of the trauma is released and the crack closes, trapping the muscles or the tissue around them in it. Entrapment of the inferior muscles can cause immediate restriction of the globe, depending upon which muscles were caught. The patient's head turns to the direction where the eye can not look freely.

Recommended treatment is to wait one to two weeks and then, if necessary, surgically correct the fracture and release the EOMs. The floor may have a gap so that the orbital contents can sink down causing *enophthalmos* (shrinking in of the globe), and a pseudo hypodeviation. Cosmetic surgical correction of the floor is then necessary.

Adherence of the EOMs to each other or to other tissue causes restrictive strabismus. Ductions are diminished and forced ductions are positive. Classically, *adherence of the EOMs* may be traced through families and usually involves all EOMs and the levator muscle, resulting in a chin-up position (Parks, 1975).

Muscle contractures, or tightening of the EOMs, occurs with any prolonged disuse of the muscle (muscle atrophy). Long-standing muscle palsies and certain syndromes (see Esodeviations: strabismus fixus) may have positive forced ductions because of tight antagonistic muscles, and not necessarily because of the palsied muscle.

Esodeviations

Esodeviations are eye turns where the eye turns inward. The main causes of esodeviations are physiologic factors, accommodative factors, and innervational factors.

Types of Esodeviations

Pseudostrabismus is a phenomenon observable in many babies. All babies DO NOT have crossed eyes. The eyes of many babies are not well coordinated until they reach six weeks of age, but any eye turn *after* that age is *not* considered normal. Many babies *appear* as though they have crossed eyes. Usually, this is due to epicanthal folds, which

Epicanthal folds make eyes appear **more crossed**. Therefore, truly straight eyes may appear ET, and truly XT eyes may appear **straight**.

give the impression that the eye is crossed inward. If the epicanthal folds are pinched up or the corneal light reflexes of the two eyes are compared, it becomes obvious the child has no eye turn. A cover-uncover test would show no eye movement. Rule out LR dysfunction by making each eye ABDuct, and consider pseudostrabismus and epicanthal folds even in an older child whose eyes don't move on the cover-uncover test. Of course remember, a child with epicanthal folds can always have a *real* ET.

Infantile and congenital esotropia refer to the same entity. Using the term "infantile" is most appropriate however—unless an ophthalmologist was present at birth to document that the ET was actually present at that time (truly "congenital").

An infantile ET is usually characterized as a large ET (> 40 P.D.) present within the first six months of life without a significant accommodative component (hyperopic refractive error). There is usually NO amblyopia because alternate fixation is a common finding with infantile ET. (see Figure 3.5). Many infants who alternate in this way **cross-fixate** or use their *right* eye to look in their *left* field over to the midline, and their *left* eye to look in their *right* field. The child avoids ABDucting either eye past the midline. For this reason, infantile ETs must be proven able to ABDuct each eye so that LR dysfunction can be ruled out.

Another common finding in infantile ETs are V patterns with overactive IO and/or dissociated vertical deviations (DVD). It is imperative to differentiate overactive IO from DVD when surgical correction of the vertical anomaly is being considered.

When one eye is occluded, **latent nystagmus** may occur where the uncovered eye beats away from the eye with the

Charactertistics of infantile/congenital ET:
1. large ET
2. present before 6 months
3. insignificant refractive error
4. cross fixation (no amblyopia)
5. V pattern with overactive IO
6. DVD
7. latent nystagmus

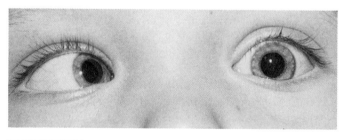

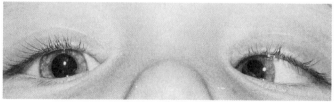

Figure 3.5 Alternating infantile ET: (top) RET. (bottom) LET.

cover. This is significant in that traditional opaque occlusion cannot be used for vision testing, or for amblyopia treatment (see Chapter 4 for vision testing techniques). Prior to surgery at a young age, vision should be equal and ABDuction well documented. Alternate occlusion prior to surgery helps achieve both of these goals. Before any type of corrective surgery, a cycloplegic refraction must be done to rule out an accommodative component.

Accommodative esotropias are classified by how much of the deviation disappears after wearing and *relaxing* into the hyperopic correction. The ET may be *fully accommodative, partially accommodative,* or *non-accommodative.* (The ET does not vary with correction). The hyperopic correction usually ranges from 2–3 D up to 7–8 D. While most children with uncorrected hyperopia are able to accommodate without converging (they use relative accommodation), some children use unwanted accommodative convergence when they accommodate to see clearly.

Classically, the time of onset is between the ages of eighteen months and three years, and begins as an intermittent eye turn occuring when the child concentrates on something. Although rare, cases of accommodative ETs manifesting prior to the age of one year do exist. Therefore, accommodative ET (hyperopia causing the eye turn) must be ruled out in children under the age of one year with ETs.

Accommodative ET also can start at a much later time (between seven and fifteen years of age). Usually, this is precipitated by some factor responsible for breaking down the previously successful use of relative accommodation in a latent hyperope (physical or emotional trauma, illness, etc.).

As the eyes accommodate on a near fixation point, they converge a fixed amount (accommodative convergence). Each person has a fixed amount of accommodative convergence per unit of accommodation: the AC/A ratio. An AC/A ratio of 5/1 indicates that for every 1 D of accommodation, there is 5 P.D. of accommodative convergence. Additional -1.00 D lenses would force the person to accommodate 1 D to see clearly through them. That person should measure 5 P.D. more convergent.

It is usually the same eye that crosses inward, so amblyopia is common. The decreased vision is an obstacle to fusion despite the straight eyes. Patching treatment for amblyopia is necessary, but decreases the amount of time spent binocularly, and thus decreases potential fusing time.

Treatment of accommodative ET is aimed at re-establishing fusion with an appropriate eyeglass correction. A low hyperopic correction is recommended to a slightly eso patient, but withheld from an exo patient if vision is not significantly affected. Similarly, a small myopic correction

Accommodative ETs with relatively lower refractive errors, (up to 4 D) are more likely to have high AC/A ratios. Accommodative ETs with high refractive errors, (over 4 D) are more likely to have normal AC/A ratios.

EXAMPLE: AC/A = 10/1, Refractive error = + 2.00 sphere, o.u. When **uncorrected,** the patient can be expected to accommodate 2 D at distance, 10/1 = ?/2 and therefore cross in 20 P.D. at distance. At near, he will accommodate 3 D more for a total of 5 D, 10/1 = ?/5, and therefore cross in 50 P.D. at near. With + 2.00 spheres on, the distance deviation should decrease to orthophoria but he will still need to accommodate 3 D at near, 10/1 = ?/3, so he will converge 30 P.D. at near with the distance correction on. When the near ET is due to the AC/A ratio, more plus correction is needed at near, so a **bifocal add** is given. It is NOT given for near vision as for a presbyope, but for correcting the near deviation; to treat the high AC/A. Because the AC/A equals 10/1, for every 1 diopter of accommodation that you let him **relax,** 10 P.D. of convergence will disappear.

With + 1.00 adds you would expect him to measure 20 ET'.

With + 1.50 adds you would expect him to measure 15 ET'.

With + 2.00 adds you would expect him to measure 10 ET'.

With + 3.00 adds you would expect him to measure ortho'.

Table 3.4 Types of Accommodative ETs

Type of Accommodative ET:	Example
Fully accommodative, normal AC/A	P + C, cc: 3 E + 5 E'
	P + C, sc: 20 ET + 20 ET'
Fully accommodative, high AC/A	P + C, cc: 5 E + 25 RET'
	P + C, sc: 25 RET + 45 RET'
Partially accommodative, normal AC/A	P + C, cc: 15 LET + 15 LET'
	P + C, sc: 25 LET + 25 LET'
Partially accommodative, high AC/A	P + C, cc: 20 LET + 40 LET'
	P + C, sc: 35 LET + 55 LET'
Non-Accommodative, normal AC/A	P + C, cc: 25 RET + 25 RET'
	P + C, sc: 25 RET + 25 RET'
Non-Accommodative, high AC/A	P + C, cc: 15 ET + 30 ET'
	P + C, sc: 15 ET + 30 ET'
High AC/A ratio	P + C, sc: ortho + 25 RET'
High AC/A ratio	P + C, sc: 20 X + 5 X'

might be withheld from a very young eso patient but encouraged for an exo. A small pre-presbyopic exo will become a larger presbyopic exo once committed to plus lenses to see clearly.

A very frequent finding in patients with accommodative ET is an AC/A ratio abnormality. If the near deviation is more convergent than the distance deviation, there is a high AC/A. The high AC/A ratio causes a disproportionate amount of accommodative convergence to accompany the accommodation necessary to see at near fixation. The eye crosses way inward for near fixation. How much it will cross inward can be predicted if the AC/A ratio is known.

Specific treatment of accommodative ET with and without high AC/A ratios will be covered in Chapter 8. Table 3.4 shows examples of the different classifications of accommodative esotropias with and without high AC/A ratios.

Monofixational Syndrome has also been called many different names but is most commonly referred to as stated; the patient is a "monofixator." It is usually an esodeviation, but it can be manifest as an exo-, hyper- or hypodeviation. It may even be manifest in an ortho patient. The affected eye turns never exceeds 8 P.D. of either ET or XT, nor more than 3 P.D. of hyper- or hypotropia.

This syndrome is characterized by mild amblyopia—as little as a ½-line difference—and by *foveal/central* suppression of the affected eye on fusion tests such as the stereotest, Worth four dot, or the four diopter base-out test.

This eye turn is obviously *NOT* noticed by casual observation and goes undetected by patients, parents, and casual examiners. Ordinarily, there is a larger deviation that can be

Monofixational Syndrome:
1. small manifest deviation (<8 P.D.)
2. larger latent deviation during cross-covering
3. mild amblyopia
4. foveal suppression
5. central ARC/Peripheral NRC

measured by prism and cover and this is often the tipoff to the diagnosis. For example, during a cover-uncover test, a very small tropia (< 8 P.D.) is noted. While measuring this tropia, the deviation builds larger and larger (up to 30 P.D.) When this happens, go back and look more carefully for a small constant tropia and for the amblyopia and suppression of the tropic eye to make the diagnosis.

While a patient with monofixational syndrome does not have bifoveal fusion, some gross fusion is present with a cosmetically acceptable eye turn. Most monofixators require no treatment except if significant amblyopia is found in a child of treatable age. Many monofixators are anisometropic amblyopes.

Duane's classification of horizontal strabismus is based on comparing near and distance deviations and attributing them to near (MR) dysfunction or distance (LR) dysfunction. The MR are more responsible at near and for convergence; the LR are more responsible at distance and for divergence. Whatever the etiology, those muscles should be directly corrected by surgery.

A **convergence excess** type deviation would be an eso that is greater at near than at distance due to MR "overaction." A **divergence insufficiency** would be an esodeviation that is greater at distance than at near due to LR "underaction."

Acquired esotropia refers to any ET that is NOT present at birth. When an older child develops a new ET, whether accommodative or due to an innervational abnormality, it is termed "acquired ET." Presuming that the child was truly ortho and fusing prior to the onset of the acquired ET, treatment is aimed at restoring fusion before any adaptations to strabismus occur.

Consecutive esotropia is an ET that is acquired postoperatively. Exo patients who undergo surgery normally have a small esotropia immediately afterward. Some esotropias may persist past the first few weeks, however, and require further treatment to eliminate the accompanying diplopia and to restore fusion.

A **CN VI palsy** causes a LR palsy where the LR function decreases while its antagonist, the MR, remains strong. This results in the eye crossing inward. The deviation will vary depending on gaze direction; it is an incomitant ET. The ET is worst where the LR usually works most: in the field of action of the LR (ABDuction). The deviation may disappear completely in the direction of gaze where the LR usually works least (so its action is not missed in ADDuction). This is where patients will most likely want to position their eyes. To do so, they must turn their *head* towards the field of the palsied muscle (see Figure 3.6).

CN VI palsies are either congenital or acquired. An acquired CN VI palsy will demonstrate incomitance and a primary and secondary deviation. The deviation will be greatest when the paretic eye is fixing (see Chapter 1). Older patients will report diplopia, and patients with a CN VI palsy may have to turn their heads in order to fuse.

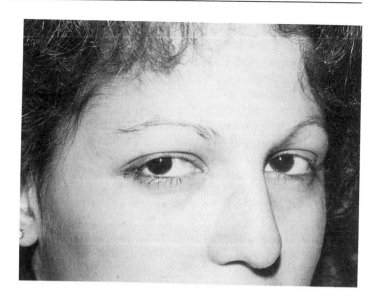

Figure 3.6 Left face turn with fusion due to **left** CN VI palsy.

Electromyography (EMG) testing is specialized testing of the actual activity of the eye muscles themselves. The muscle activity is recorded from needle probes which are placed directly into the anesthetized EOM. The activity is recorded as the awake patient is commanded to look in various positions of gaze.

CN VI palsy is frequently caused by head trauma, neoplasm (in children), or ischemic causes (in adults). It is rarely caused by aneurysm, and approximately 8–30% are of an undetermined origin (Bajandas, 1980). Because of the implications of a VI palsy, children with an ET must demonstrate good ABDucting ability of the eyes for LR dysfunction to be ruled out. Other testing methods, such as saccadic velocity testing and electromyography (EMG) testing, show a decrease in LR activity.

Cyclic esotropia is an unusual phenomenon that can be diagnosed by history. A young child sometimes develops a 24-, 48-, or 96-hour cycle where the eyes alternate between being normal (straight and fusing), to having a large ET with suppression and possibly ARC. The child should be measured on a straight day, followed by measurements in the cycle on a crossed day (see Figure 3.7). Because the child has a large ET 50% of the time, suppression, ARC, and amblyopia may develop. The amblyopia should be treated, and surgery should be performed in accordance with the amount of

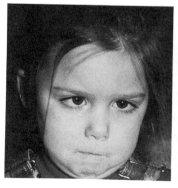

Figure 3.7 Cyclic ET: (Left) RET on crossed day. (Right) five days later (five 24-hour cycles) orthophoric.

esotropia present on crossed days. Ordinarily, the cycle breaks down so that the child has a large ET every day. However, surgery can be done without consequence before this decompensation occurs (von Noorden, 1985).

Divergence paralysis may be caused by a variety of conditions including encephalitis, increased intracranial pressure, or trauma (von Noorden, 1985). It is characterized by an ET that is greatest at distance because of the eyes' inability to diverge. There is little or no deviation at near because divergence is not exerted at near. Because divergence paralysis is *NOT* due to LR dysfunction specifically, the ET is *comitant* and does not vary significantly in right or left gaze. Versions and ductions may be normal but fusional divergence is negligible. Divergence paralysis most frequently occurs in visually mature individuals and therefore causes horizontal, uncrossed diplopia—which would be the patient's main complaint. The diplopia is relieved by occlusion or base-out prism. The prisms can be gradually reduced and, in some extreme cases, bilateral LR resections may be necessary.

Divergence paralysis may mimic bilateral LR palsies. Measurements will be **comitant** (the same in all directions of gaze) in a patient with divergence paralysis.

Strabismus fixus is a large ET where both eyes are fixed in the extreme ADDucted position. Neither eye can be forced into ABDuction because of extreme muscle contractures. It can be caused by any long-standing ET, where the MR have become so contracted and inelastic that the eye cannot be physically rotated into ABDuction. There are positive forced ductions. The ADDucted eye position necessitates a head position for cross-fixation. Treatment is aimed at facilitating fixation without necessitating the extreme head position.

Syndromes

Duane's retraction syndrome type 1 is an esotropia present at birth because of misfiring of the MR and LR. In primary position, a small ET is often present, but the patient may be ortho. When ABDuction of the affected eye is attempted no innervation goes to the LR so the eye does not go past the midline into ABDuction, and the eyelid fissure widens. In ADDuction, both the MR and LR are innervated simultaneously (not in accordance with Sherrington's law of reciprocal innervation) and the eye pulls back into the orbit, causing narrowing of the palpebral fissure. ADDuction is often slightly deficient. Because fusion can often be attained without significant head positioning, surgery is usually not necessary. There is the risk of amblyopia developing and this should be treated appropriately. My best advice is to encourage the use of head positioning.

Nystagmus compensation syndrome (NCS) causes a large ET and becomes apparent shortly after birth. It is thought to be caused by jerk nystagmus, present only with the eye in ABDuction. To compensate so vision is better, both eyes are

Nystagmus Compensation Syndrome (NCS) may mimic infantile ET. The presence of nystagmus when ABDuction is attempted rules out infantile ET.

Syndromes involve two or more findings that consistently go together (i.e. Horner's syndrome involves ptosis, miosis and anhydrosis).

kept in ADDuction. Since neither eye can be out of ADDuction *without* nystagmus, cross-fixation with a head turn results. NCS looks similar to infantile ET, except jerk nystagmus is present when either eye gets past the midline into ABDuction. Parents frequently have never seen the nystagmus.

The tip-off in discovering a child has NCS occurs when quantifying the eye turn by Krimsky measurements. The correcting base-out prism neutralizes the manifest deviation, and it forces one eye to the midline. Immediately, nystagmus is evident or the child over-converges *again* and a large ET still exists, despite the "correcting" base-out prism. Treatment is surgical (Frank, 1979).

Treatment of accommodative effort syndrome: Plus correction is given to diminish the need for accommodation and provide clear near vision.

Accommodative effort syndrome is a rare phenomenon resulting in a small esophoria at near and asthenopic symptoms for reading. The monocular NPA is normal, but accommodation is reduced when measured binocularly. There are poor amplitudes of relative fusional divergence but, as would be expected, increased amplitudes of accommodative convergence. Blurred vision during near fixation is caused by the patient's effort to have comfortable single binocular vision. The blurring is caused by *relaxed* accommodation, which decreases the accompanying accommodative convergence (Hurtt, Rasicovici, and Windsor, 1977).

Gradenigo syndrome is a unilateral CN VI palsy caused by a middle ear infection. Pain is experienced on the affected side of the face, possibly accompanied by partial or total deafness. Usually, the CN VI and ET are spontaneously resolved with treatment of the ear infection. During the esotropia, amblyopia and MR contractures may be prevented with occlusion (Harley, 1975).

Moebius syndrome is a bilateral congenital syndrome in which a CN VI and a CN VII palsy combine, causing an ET and an inability to form facial expressions. An inability to completely close the eyelids may also result. The ET requires surgical treatment aimed at moving at least one eye to the primary position, thus enabling the patient to fixate without a large head turn.

Exodeviations

Types of Exodeviations

Pseudo exotropia is most often caused by a nasally located pupillary reflex. The corneal light reflex in a person's eye does not usually fall directly in the center of the pupil. This is because the visual axis of the eye does not quite coincide with the geometric axis. The angle between the visual axis and the geometric axis is referred to as **angle kappa** although it is technically angle lambda (von Noorden, 1985). A **positive angle kappa** (nasally located reflex) simulates an exodeviation because, with a nasally located corneal light reflex, the eye appears to have turned outward; a pseudo XT exists. Only a cover-uncover test accurately reveals a true tropia. Asymmetrical angle kappas are most deceiving. Angle kappas are measured in degrees by a haploscopic instrument, but are best documented with photographs of monocular fixation.

The majority of the population has a slightly nasally located light reflex, a positive angle kappa, and up to 5° is considered normal.

Congenital XT is rare, and an exotropia of any kind in an infant requires thorough investigation. As stated before, eye turns in infants are NOT normal and should be investigated by an ophthalmologist. When decreased vision or disease is eliminated, the infant is considered to have congenital/infantile exotropia.

XT in an infant can be caused by poor vision in the nonpreferred eye by a number of vision and/or life threatening diseases: retinal detachment, vitreous/retinal hemorrhage, PHPV (persistent hyperplastic primary vitreous), retinoblastoma, congenital cataract, or CN III palsy, or midbrain dysfunction.

Congenital XT is characterized by a large, constant XT with a strongly preferred eye and amblyopia of the other. The amblyopia is treated by routine methods, but perfect binocular vision is unlikely. Once the vision is equal, the cosmesis can be treated surgically only. Although amblyopia is the rule, patients occasionally have equal vision and do the opposite of cross-fixation. Congenital XT patients may use their ABDucting eye to fixate and be reluctant to ADDuct either eye. When surgery is performed on an adult or an older child with a long-standing XT, the patient sometimes experiences a subjective loss of peripheral vision postoperatively.

General Exodeviations

The vast majority of exo patients are at some stage between being purely exophoric and being mostly exotropic. Many exo patients start as phorias, then decompensate into intermittent tropias and constant XTs. Throughout these changes, the *size* of the deviation does not vary much; most are between 20–30 P.D. Most also exhibit some degree of bicentral fusion, and have nearly equal vision.

From the patient's history, it must be determined if decompensation is occurring. How much time an X(T) spends in the "phoric" state in comparison to how much time he spends in the "tropic" state must be noted. It is also necessary to determine if the XT is getting worse (decompensating) or staying the same. Be more specific than just asking: "Is the eye turn getting worse?" You have to ask if the amount of times daily that the eye deviates is more or less, and when it does deviate, does it stay XT for longer periods of time? Someone who is straight and fusing 90% of the time is far better off than someone who is tropic and adapting to that strabismus most of the time.

Decompensating XTs require treatment intervention so that suppression, amblyopia, and ARC do not develop. The goal of treatment in a decompensating XT is to regain binocular fusion.

Tropic time is most damaging to a young child because

adaptations to strabismus occur most rapidly and easily. Suppression will be reinforced each *minute* that is spent tropic. Amblyopia occurs if there is a preferred eye, and ARC is also more likely to occur. When patients in the tropic state blink or change fixation (or are badgered enough by their mothers), they convert back to the phoric state and have straight eyes and excellent fusion. Very few patients ever report that they "know" when they go tropic because of the loss of depth perception, but some people think they can "feel" the eye go. Most have to rely on a faithful observer (mom) to bring the XT to their attention. Tell mom that while nagging is a terrible thing, it's worth it in this case. The child must spend the most minutes possible each day keeping the eyes straight and reinforcing fusion.

Table 3.5 shows the different adaptive changes characteristic of exophorias, intermittent exotropias, and constant exotropias. A patient with an *exophoria* never suppresses and always maintains bifoveal fusion, provided he has adequate convergence amplitudes. Remember that a phoria only manifests itself when fusion is disrupted, as with a cover-uncover test. With the eyes in the ortho position, the exophoric patient has NRC and good fusion.

An exophoric patient who suppresses his temporal retina when bifoveally fixating will not appreciate diplopia if he slips exo. One moment he could be straight and bifoveally fixating; the next moment he could be XT and suppressing with ARC. This is the *intermittent exotropia*. Awareness of diplopia makes an XT aware of diplopia when the eyes slip tropic. Teaching diplopia awareness is the goal in treating patients with decompensating exodeviations who have fusion potential.

Constant XTs are like X(T)s in their tropic state; they suppress and often have ARC. These patients NEVER pull straight and fuse. Constant XTs represent a real failure in treatment if they were "allowed" to decompensate from a phoria or X(T) into this state. Treatment intervention must take place before fusion is lost completely and cannot be regained. If treatment—either surgical or therapeutical—occurs before fusion is lost, fusion is often maintained and even reinforced.

When examining a patient with a long-standing, constant XT, try stereo testing. Patients with even very large devia-

Table 3.5 Exodeviations

	Fusion	Suppression	?ARC
Exophoria	yes	no	no
Intermittent XT	yes	yes	maybe when tropic
Exotropia	no	yes	usually

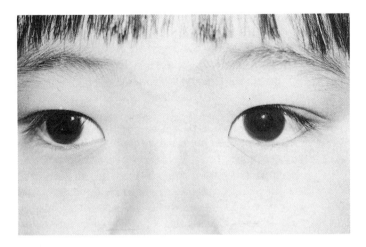

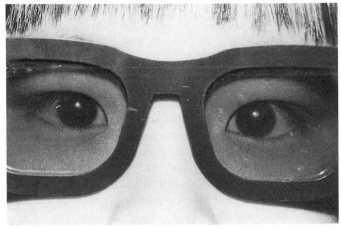

Figure 3.8 (Top) patient with large RXT. (Bottom) pulls straight and fuses with polaroid glasses on for stereotesting.

tions (70–90 P.D. XT) sometimes pull straight and have perfect fusion during stereo testing (see Figure 3.8)! For the remainder of the exam—as with everyday seeing conditions,—they have a constant XT, suppression, and no fusion.

As mentioned previously, most exodeviations fall somewhere between being a phoria and being an X(T). Many of them have a very small comitant hyperdeviation (1–2 P.D.) and most have a slightly preferred eye, perhaps with mild amblyopia. Treatment is aimed at achieving comfortable, single binocular vision with fusion. Either the deviation is minimized by surgery and/or fusional convergence amplitudes are improved with convergence training.

Duane's Classification of Horizontal Strabismus

An exodeviation that is larger at near than at distance would be called a **convergence insufficiency** type of exodeviation

Duane's Classification of Horizontal Strabismus also includes exodeviations. The classifications are based on **size**, not control (i.e., phoria or tropia). Any deviation that is the same measurement at distance and near is a **basic deviation**.

and the MR would be considered "underactive." This type of exodeviation is not to be confused with the malady called "convergence insufficiency," where the patient has asthenopic symptoms with prolonged visual use. "Convergence insufficiency" is due to insufficient *fusional convergence* for the person's visual task at distance and/or at near, and *may or may not be associated with an eye turn.*

An exodeviation that is larger at distance than at near would be termed a *divergence excess* and the LR would be considered "overactive" (see Table 3.6).

Either because the fusion reflex is so strong in some XTs or because the AC/A ratio is high, some apparent divergence excess type XTs are really only **pseudo divergence excess.** Merely doing a cover-uncover test,—or even a cross-cover test—does not disrupt near fusion significantly, and patients appear to have little or no near exodeviation. Prolonged occlusion (24 hours) is necessary to uncover what is actually a basic deviation: the near equals the distance deviation. The patient is to wear a patch full-time for 24 hours before *you* occlude the open eye, remove the patch, and do prism and cross-cover measurements (see Figure 2.11). The patient with pseudo divergence excess is phoric. The deviation can't possibly be intermittent or constant because extreme measures (a 24-hour occlusion) are necessary to bring out the deviation. If, after 24 hours of occlusion, the near deviation is still smaller than the distance deviation, the strabismus is classified as a **true divergence excess.** If the near deviation becomes the same as the distance deviation, it is a *pseudo divergence excess.* Patching is the only way to determine this type of strabismus. (Measuring with plus lenses—as sometimes advocated—only measures the gradient AC/A ratio.)

Other Causes of Exodeviations

Isolated **MR palsy** (or CN III palsy) causes an exodeviation because the eye cannot ADDuct well. The deviation will be greatest with the paretic eye fixing (the secondary deviation) and in the paretic MR's field of action (ADDuction). An abnormal head position may be adopted to yield fusion or facilitate fixation (see Figure 3.9).

Uniocular blindness in a visually mature patient who

Table 3.6 Duane's Classification of Horizontal Strabismus

1. Basic deviation	ET or XT, D = N
2. Convergence excess	ET, N > D, MR "overactive"
3. Divergence insufficiency	ET, D > N, LR "underactive"
4. Divergence excess	XT, D > N, LR "overactive"
5. Convergence insufficiency	XT, N > D, MR "underactive"
6. Pseudo convergence insufficiency	XT, D = N with patching

originally had equal vision and binocular fusion results in a loss of that fusion. This may cause the eye to develop an exodeviation. Patients with a cataract in one eye often have exotropias that persist after equal vision is restored with surgery, an intraocular lens, or a contact lens. Vision is good, and the image size discrepancy (aniseikonia) is minimized but the exotropia is too large for fusion to occur. You should advise these patients to WAIT, and *discourage temporary occlusion to eliminate the diplopia*. It may take months, but they often miraculously regain fusion on their own.

When the fusion mechanism weakens in a previously fusing patient, it can result in an exodeviation. If an older patient develops a "new" exodeviation (which may or may not be accompanied by diplopia), try to determine if the patient was previously fusing, and eliminate all other causes of a "recent onset" exodeviation.

Isolated MR palsy is uncommon, so a pure exodeviation would usually not be the presenting sign of a CN III palsy. Myasthenia gravis or internuclear ophthalmoplegia may mimic an isolated MR palsy.

Syndromes

Duane's retraction syndrome type 2 is similar to type 1 in that there is globe retraction and lid fissure narrowing when

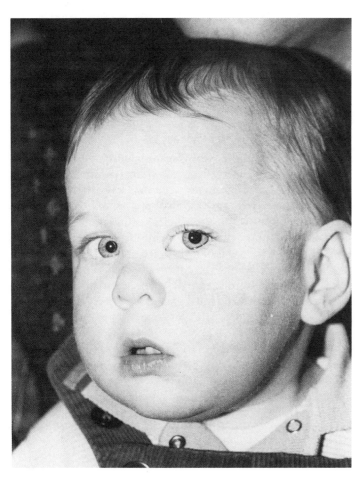

Figure 3.9 Right face turn to yield orthophoria in child with left MR palsy. (LMR works best in **right** gaze.)

ADDuction is attempted. Type 2 differs from type 1 in that ADDuction decreases and ABDuction is unaffected. This results in an exodeviation – when the affected eye attempts ADDuction. There may be a small exo in the primary position, or the patient may be ortho. Type 2 is best documented by photographs of the eye movements. It usually does not require treatment unless amblyopia develops, or if cosmesis is objectionable.

> Duane's Syndrome Type 2 may mimic an isolated MR palsy. Globe retraction and lid fissure narrowing when ADDuction is attempted would indicate Duane's Syndrome Type 2.

Benedikt's syndrome is caused by a lesion in the brain stem near the CN III nucleus (a lesion of the red nucleus). This, in turn, causes a CN III palsy and hemiparesis (half-body paresis). A lesion on the right side of the brain causes the CN III palsy to be on the right and the hemiparesis to be on the left.

AC/A Ratio and Exodeviations

AC/A ratio also affects exodeviations. Plus lenses relax accommodation and, therefore, relax accommodative convergence. All emmetropised patients can anticipate becoming presbyopic during their 40's to 50's. A pre-presbyopic patient who has convergence insufficiency, an exodeviation, or a high AC/A ratio may experience symptoms related to increased fusional convergence requirements when his first near adds are prescribed. Although the print on the page is clear, the plus lenses cause a larger exodeviation and, therefore, an increased convergence requirement. This warrants fusional convergence training, and in some cases of very large exodeviations, surgery is necessary. Patients are often very unhappy because they were told their original presbyopic symptoms would vanish with the new glasses. The new glasses however, made their symptoms worse in a different way. *Predicting* the patients to whom this will happen and advising them is not always possible. Therefore, all patients with new plus prescriptions, regardless of whether the prescription is a near add or a distance correction, should be advised. Patients should have their exodeviation and fusional convergence measured with the new plus lenses *before* the lenses are prescribed. Any increase in exodeviation or decrease in fusional convergence warrants fusional convergence training or, at the very least, a good warning to the patient.

> Exo patients with high AC/A ratios who become presbyopic, may experience asthenopic symptoms due to their exodeviation increasing while wearing their new plus lenses. Good fusional convergence amplitudes are necessary to comfortably control an exodeviation.

Plus lenses relax accommodation and accommodative convergence, which would increase an exodeviation. Minus lenses stimulate accommodation and accommodative convergence, and would decrease an exodeviation. A typical four-year-old does not need to have a balanced –0.75 sphere myopic refractive error corrected. His need for clear vision at distance does not warrant wearing (and breaking, and losing, and scratching) glasses. *If* that child is also significantly exo, however, minus lenses may control the exodeviation so that impending decompensation is arrested.

In the same way, a typical three-year-old with a balanced + 2.50 sphere refractive error, no symptoms of decreased vision, and no eye turn would not be given glasses. The same three-year-old with an accommodative esotropia *would* be given the glasses to help control the esodeviation. If that child had an exodeviation, the correction would NOT be given — even if vision were sacrificed — because maintenance of fusion was considered temporarily more important than the vision.

Somehow, withholding a hyperopic correction from a young child with an exodeviation is different from prescribing over-minusing lenses to an emmetropic child with an exo-deviation. In both cases, however, stimulation of accommodation and accommodative convergence would probably decrease the exodeviation, making it easier for the child to maintain fusion.

Many young children are hyperopic, have no symptoms of decreased vision or eye turn, have never been examined by an ophthalmologist, and walk around without corrections. Prescribing over-minused lenses tamper with an emmetropic patient's refractive state. The lenses are expensive (compared to simply withholding a hyperopic Rx), and it is difficult to keep the glasses on the child. The glasses don't make the child see better, but do increase the accommodative strain. An uncorrected hyperope is more accustomed to accommodative strain. The over-minused exo patient must eventually eliminate the minus lenses, and the exodeviation returns. Over-minusing doesn't really work as fair, permanent treatment of exo patients.

Vertical Deviations

Vertical deviations are usually accompanied by eso- or exodeviations. They may be restrictive and nearly all are incomitant in nature.

When a vertical deviation that does not seem to be restrictive is detected, a palsied muscle may be present. Muscle palsies may have various etiologies: congenital, traumatic, or compressive lesions on nerves, or systemic myasthenia gravis. It is important to isolate the muscle at fault and to document the eye turn with measurements in all fields of gaze.

The **Bielschowsky three step test (B3ST)** is only accurate in a situation where there has been no previous surgery and

The **Bielschowsky Three Step Test** (B3ST) helps isolate which of the eight cyclovertical muscles may be at fault in a vertical muscle palsy.

where there are no restrictions of the EOMs. In a long-standing muscle palsy there has frequently been a nearly complete spread of comitance and there may not be much variation in the deviation to right and left gaze.

Longstanding Muscle Palsies

Long-standing muscle palsies that were present at age six or earlier have certain characteristics that recent onset palsies do not. Although people naturally exhibit horizontal fusional amplitude, they have very little vertical fusional amplitude and can overcome only 2–3 P.D. of vertical deviation maximally. This explains why thyroid patients with very small hypertropias have very bad diplopia. In a congenital/long-standing vertical palsy, the patient frequently develops vertical amplitudes and can control the vertical deviation; the palsy remains phoric. (It may measure up to 40 P.D.)

When measuring vertical amplitude, record the point where diplopia occurs and the recovery point where fusion can be re-established. For instance, if the patient broke into diplopia at 7 BD, and then recovered at 6 BD, record it as "BD OD: $^7/_6$." Repeat measuring vertical amplitude using BD OS and record the break and recovery points.

The presence of *vertical amplitudes* at distance, at near, or both strongly suggests the palsy was present at, or before, age six. Use a vertical prism bar starting with 1 P.D. base down (BD) and progress to greater and greater prism. First, measure base down OD (see Figure 3.10), which causes the right eye to elevate in relation to the left to maintain fusion; a person with a RHT from a congenital/long-standing muscle palsy would exhibit this. Whichever eye was able to overcome the *most* BD prism should be the eye with the hyperphoria/tropia. Any asymmetry of the eyes (i.e. BD OD: 7/6, BD OS: 3/2) is considered significant.

Patients with congenital/long-standing cyclovertical muscle palsies do not perceive torsional diplopia as a patient with

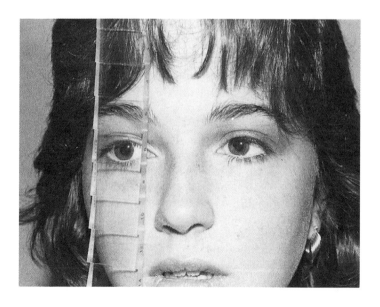

Figure 3.10 Vertical amplitudes are measured with a vertical prism bar measuring base down OD first (shown), then base down OS.

a recent onset palsy would. Although the eye itself may tort, torsion is not perceived; it is as though the patient had developed torsional ARC. To document this (and any torsion actually perceived by the patient), use the **double Maddox rod test.** Two Maddox rods, a red and a white, are set in a trial frame in a completely darkened room. Set both at 90° so the patient will see the red-and-white lines horizontally. An ortho person will perfectly superimpose them. A person with a HT and NRC will see the appropriate vertical diplopia. (When the red Maddox rod is placed over the right eye [with a RHT], the *white* line is seen above the red line; see Chapter 6.) Someone with torsion and NRC will see one line parallel to the floor and one line tilted. Because monocular image projection is opposite to objective location when the eye is intorted (i.e. if the eye is in, the object is out), the image seen is extorted. My advice is to stand *behind* patients when testing with double Maddox rods, and have them indicate with their arm how the line is tilted (see Figure 3.11). An intorted image is caused by extorsion of the eye, so an intorter muscle (SR, SR) is at fault. An extorted image results from intorsion of the eye, with an extorter muscle (IO, IR) being weak.

A patient with a recent onset cyclovertical muscle palsy will appreciate torsion. Torsion may be approximately measured by rotating the Maddox rod in the trial frame so as to make the red and white lines parallel to each other. The degrees of torsion may then be estimated from the trial frame. Someone with a long-standing cyclovertical muscle palsy would not perceive any torsion (Guyton, 1983). Guyton describes the use of indirect ophthalmoscopy to estimate objective torsion.

A patient with a **recent onset cyclovertical muscle palsy** will appreciate torsion but will **not** have any vertical amplitudes.

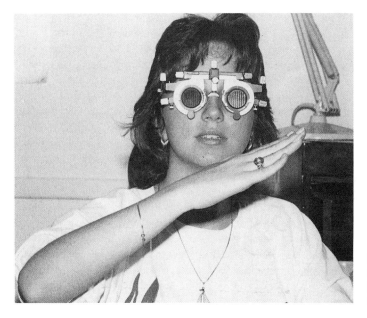

Figure 3.11 Patient indicating direction of tilted line during double Maddox rod testing for torsion.

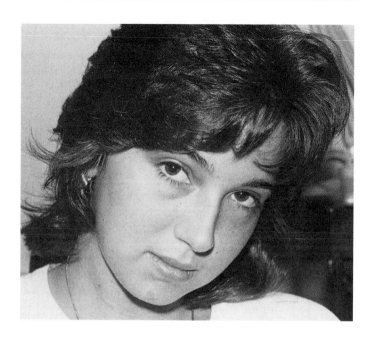

Figure 3.12 Head position with RSO palsy: chin down, face left, tilt left.

Types of Vertical Deviations

Paralytic vertical deviations are caused by CN palsies involving one or more of the cyclovertical muscles. They are characterized by their incomitance and by the presence of a primary and secondary deviation, both of which diminish with time due to contractures of the EOMs. The B3ST helps determine which muscle is at fault (see Chapter 6). Some patients regain fusion by tilting or turning their head to put their eyes closest to the ortho position. The head is usually positioned so the nose points towards the direction of gaze in which the palsied muscle works. For instance, the LLR works in left gaze, so the head would turn to the left for viewing things in right gaze. The RSO works best in down- and left-gaze, so the head will point in this direction to allow for viewing in the opposite field (see Figure 3.12).

Correcting prism may help alleviate symptoms of diplopia and asthenopia in a fairly comitant deviation. When prisms fail, surgical correction may be necessary. Surgery is performed to restore single binocular vision to some useful field of gaze, usually primary position and the reading position.

DEP is thought to possibly be a "supranuclear palsy" because no single lesion in the brainstem could explain a SR palsy and an IO palsy without any other findings.

Double elevator palsy (DEP) is a congenital malady that renders the eye completely incapable of elevation. A child's eye will not elevate to the midline and is hypotropic in primary position. The child may fuse in downgaze, thus having a chin-up head position to allow this. If the child prefers fixation with the affected hypotropic eye, a chin-up head position is necessary for seeing. The lid *appears* ptotic due to the hypotropia (pseudo ptosis). Amblyopia may develop, requiring patching therapy. Forced ductions should be negative, but may indicate inferior muscle restrictions in a longstanding DEP where contractures of those muscles from disuse may have occurred. Surgical correction may be necessary for cosmesis and/or to allow fusion in the primary position.

*Pseudo ptosis is due to the **hypo**tropic position of the eye. If forcing the hypotropic eye to fixate eliminates the "ptosis," it is really a **pseudo ptosis**.*

Dissociated vertical deviation (DVD) is a type of vertical deviation that defies good explanation. Either eye will elevate under cover occlusion, yet a DVD does NOT behave as a hyperphoria/tropia. In a true hypertropia, the hypotropic eye remains relative hypo once the eyes are dissociated. This holds true regardless of the position of the relatively hypertropic eye—even if that eye is fixing.

In a DVD, the eye that is NOT elevated NEVER drops below the midline. Most DVDs are asymmetrical, and may present as unilateral DVDs. The non-hypertropic eye NEVER drops below midline during cross-covering.

DVD may occur with any form of strabismus, even in conjunction with true hypertropias. It is most commonly found in children with infantile/congenital ETs. Frequently, the DVD is not immediately obvious, and may not become obvious until after the horizontal deviation has been surgically corrected. The fact that there is a high incidence of V pattern with overactive IO in infantile/congenital ETs may also prolong discovery. Overactive IO mimic DVD, and vice versa. These must be differentiated.

DVD is different from overactive IO because an eye with a DVD will elevate under cover even from the ABDucted position but will NEVER drop into a hypotropic position.

Because the eye is usually ET to begin with, it is already in ADDuction, where the IO elevates most efficiently. Extreme ADDuction in the nonpreferred eye also dissociates that eye from the other. Therefore, if there is a DVD, it becomes manifest as the ADDucting eye is occluded from fixation by the nose. In either case (DVD or overactive IO), the ADDucting eye elevates. An overactive IO in need of surgical correction and a cosmetically unacceptable DVD are different. It is imperative that the two be differentiated. (Of course, both may exist simultaneously.) There are two main ways to differentiate a DVD from an overactive IO.

First, an eye with a DVD *also* elevates under cover or while dissociated in ABDuction. This can be demonstrated by moving the suspect eye into ABDuction and then occluding it. If the eye goes up while under the cover, a DVD exists (see Figure 3.13). An overactive IO will *not* elevate in ABDuction.

Second, an overactive IO will demonstrate a hypodeviation of the opposite eye. Ordinarily, when testing versions in dextro- or levoversion, the fixing eye is the one ABDucting. This allows the non-fixing eye to elevate, either because it has DVD or because it is in the field of the overactive IO. If the ADDucted eye elevates because the IO on that side is

DVD

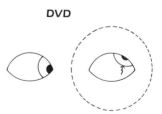

Figure 3.13 True DVD: Left eye elevates in ABDuction under cover.

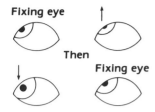

Fixing eye

Then

Fixing eye

Figure 3.14 True overactive IO: (top) When the ABDucting eye is the fixing eye, the ADDucting eye elevates. (bottom) When the ADDucting eye is forced to fixate (by occluding the ABDucting eye), the ABDucting eye drops.

Vertical deviations caused by restrictive strabismus:

1. Grave's ophthalmopathy
2. Brown's syndrome
3. Duane's retraction syndrome types 1, 2, 3

overactive, forcing the over-elevated eye to fixate will cause the ABDucting eye to become *hypo* (see Figure 3.14). ONLY an overactive IO will do this. (The eye opposite the one with a DVD would not become hypo when fixation is switched.)

Comitant Vertical Deviations

As stated previously, some long-standing paralytic deviations may have such complete spread of comitance that the *hyperdeviation* is entirely *comitant*. This type of deviation is most amenable to treatment by vertical prism correction to restore binocular vision and/or relieve diplopia. Small amounts of prism may be ground into spectacle correction and still be cosmetically acceptable. Larger amounts of prism are not cosmetically acceptable in glasses, and are often best treated by press-on Fresnel prisms. Many large eso- and exodeviations are also accompanied by very small (1-2 P.D.) hypertropias. Surgical correction of the horizontal deviation often rids the patient of the hypertropia, too.

Incomitant Vertical Deviations

Restrictive strabismus often causes vertical deviations that have already been cited. Thyroid ophthalmopathy is the main restrictive strabismus that causes primarily vertical deviations. As the muscles (particularly the IR) lose their elasticity, upgaze becomes restricted and a vertical deviation results.

Brown's SO tendon sheath syndrome prohibits the SO from moving freely through the trochlea when the eye attempts to move up and in. This causes a hypodeviation of the affected eye. Occasionally, the affected eye even has a hypotropia in the primary position.

Duane's retraction syndrome, although a horizontal syndrome, sometimes exhibits up- or down-shoots of the affected eye because of the tight LR. As the eye attempts ADDuction, the tight LR "slips" on the globe and the eye shoots up or down. This is NOT oblique dysfunction, but is due to a very tight LR slipping on the globe.

References

Bajandas, F.J. *Neuro-Ophthalmology Board Review Manual.* Thorofare, NJ: Charles B. Slack, Inc, 1980, p 71.

Bredemeyer, H.G, Bullock, K *Orthoptics Theory and Practice.* St. Louis: C.V. Mosby, 1968, p 249.

Frank, J.W. Diagnostic signs in the nystagmus compensation syndrome. *Journal of Pediatric Ophthalmology and Strabismus* 16:5, 317-320, 1979.

Guyton, D.L. Clinical assessment of ocular torsion. *American Orthoptic Journal* 33:7-15, 1983.

Harley, R.D. *Pediatric Ophthalmology.* Philadelphia: W.B. Saunders, 1975, p 458.

Hurtt, J. Rasicovici, A. Windsor, C.E. *Comprehensive Review of Orthoptics and Ocular Motility.* 2nd ed. St. Louis: C.V. Mosby, 1977, p 145.

Killiam, P. McClain, B., Lawless, O. Brown's syndrome—an unusual manifestation of rheumatoid arthritis. *Arthritis and Rheumatism* 20:5, 1977.

Parks, M.M. *Ocular Motility and Strabismus.* Hagerstown, MD: Harper and Row, 1975, p 171.

von Noorden, G.K. *Burian-von Noorden's Binocular Vision and Ocular Motility.* 3rd ed. St. Louis: C.V. Mosby, 1985.

CHAPTER 4

Adaptations to Strabismus

Introduction

The brain reacts to strabismus with protective mechanisms. Generally, the younger child is more able to cope with strabismus because the young visual system is most "pliable." That is, the visual system is immature. Also, the longer that the strabismus is manifest (either minutes spent tropic in the day or days spent since the onset of the strabismus) the more likely the adaptations have taken place.

Young children with eye turns are usually unaware they have strabismus until they look in the mirror. This unawareness is the result of the brain's adaptations to strabismus.

There are three major mechanisms for adapting to strabismus: suppression, amblyopia, and abnormal retinal correspondence (ARC).

Strabismus potentially initiates three reactions in the young child: the development of suppression, amblyopia, and ARC.

Suppression

Suppression is the first protective mechanism activated with the onset of strabismus. This occurs soon after strabismus starts, so nearly 100% of young patients with strabismus have learned to suppress.

Suppression is the shutting off of one eye during binocular viewing. It occurs only when both eyes are open, and only in the eye that is deviating at the time. As soon as the fixing eye is covered, the deviating eye stops suppressing and points its fovea at the object of regard. Suppression blocks out the images that would normally be seen by the deviated eye. If the acuity of the suppressing eye were measured, it would be very poor. By forcing fixation, suppression stops and vision returns to normal.

Confusion and Diplopia

Normally, a patient with straight eyes has a "correspondence" between the fovea of one eye and the fovea of the other. The brain knows that images from each fovea are to be superimposed and fused if they are similar enough. When a tropia occurs, the foveas no longer point at the same object. The fixing fovea sees the object of regard while the deviating fovea sees something else. Instinctively, the brain tries to superimpose these two dissimilar images and "confusion" occurs. **Confusion** is the result of corresponding retinal points being stimulated by two different objects (see Figure 4.1). The brain's adaptation to confusion is the selective mental shutoff of the deviating eye's foveal area, thus selectively ignoring the extra "confused" image. Confusion is difficult to recognize, even for astute adult observers who have a tropia induced by prism. What *is* immediately noticed (and very bothersome) is diplopia.

Diplopia is the result of one object stimulating noncorresponding retinal points (the fovea of the fixing eye and a nonfoveal point in the deviating eye). While this double vision is a tip-off to the patient that strabismus is manifest, it is a real problem if the patient lacks the vergence ability to return the eyes to the straight-ahead fusing position.

A young patient can learn to suppress the region of the retina that is responsible for the double vision under binocular conditions. The nasal retina of an ET patient's deviating eye is stimulated by the object of regard. Therefore, it is the nasal retina that "learns" suppression in an ET patient. The temporal retina of an XT patient's deviating eye is stimulated by the object of regard. Thus, the temporal retina learns suppression in the XT patient. Superior retina learns suppression in a hypertropic eye; inferior retina learns suppression in a hypotropic eye.

Confusion is rarely noticed by patients. Diplopia is obvious but is quickly suppressed in the visually immature patient.

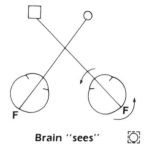

Figure 4.1 A RET results in the fovea of the right eye pointing at an object other than the object of regard. Attempting superimpositioning results in confusion.

SUPPRESSION
Esotropic eyes suppress nasal retinal. Exotropic eyes suppress temporal retina. Hypertropic eyes suppress superior retina. Hypotropic eyes suppress inferior retina.

Depth of Suppression

Suppression is not "all or none"; a patient does not either have it or not have it. Some patients have suppression under all seeing conditions and some have it only under certain seeing conditions. Certain conditions make it more likely for patients to be aware of two images; other conditions make them more likely to suppress. A patient may suppress in one position of gaze but not in another, where suppression was never learned. The "turning off" of the extra image is an instantaneous, split-second reaction.

Tropic patients are most likely to suppress (that is, be unaware of diplopia) under *normal seeing conditions* — where the room lights are on and the patient is viewing *real* objects. They are least likely to suppress under abnormal seeing conditions.

Looking at a bright light may cause patients to suppress their deviating eye. As background lights are dimmed, the object of regard appears brighter, and suppression may disappear. This occurs because the increased contrast between the

Depth of suppression may be assessed by:
1. Using a bright light for fixation
2. Dimming background room lights
3. Using red filter(s)
4. Using the Maddox rod
5. Introducing vertical prism
6. Using a haploscopic instrument

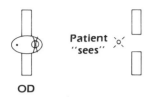

OD

Figure 4.2 Maddox rod and suppression scotoma: Image of the Maddox rod falling on and extending past the suppression scotoma of deviated right eye when the left eye was fixing on the light (left). The patient would then see a white light and the red line but the line would have a central gap missing from it (right).

Four Diopter Base-Out Test for Suppression/Fusion: The eye with suppression will not move when the prism is placed over it. If either eye moves in towards the nose at any time during testing, that eye is not suppressing.

background and the object of regard causes the light to be seen more easily through the suppression area. Dissimilar images make patients increasingly aware of diplopia because the images become increasingly dissociated. Introducing a red filter to the fixing eye causes the two images to differ in color. Using increasingly darker red filters make the differences of the images even *more* likely to be noticed. Using a vertical Maddox rod instead of a plain red filter may make the image extend beyond the suppression scotoma (see Chapter 6; Figure 4.2).

Vertical prism would also move the retinal image out of the suppression region, but would not actually break through the suppression. A haploscopic instrument such as a synoptophore can measure the depth of suppression using light intensity controls. The synoptophore can make the fixing eye's target to be *less* brightly illuminated and/or the suppressing eye's *more* brightly illuminated. Documenting the seeing condition at which suppression is replaced by diplopia may be helpful in preoperative patients, or when treatment of suppression is being considered.

Four Diopter Base-Out Prism Test

A suppression response to any of the aforementioned tests in the presence of an obvious tropia is easily interpreted. Testing for suppression in presumably straight eyes is more difficult. The **four diopter base-out test** checks for the presence of fusion or suppression. A "negative" four diopter BO test indicates fusion; a "positive" test indicates suppression.

When the eyes are straight and fusing, a small base-out prism introduced to either eye will ultimately cause a vergence movement. This is because the *base-out* prism induced a small exodeviation (with crossed diplopia), so a small *convergence* movement will be necessary to bifoveally fixate again. A small (4 P.D.) prism is used because almost all fusing patients are able to converge four prism diopters.

A patient who foveally suppresses will not be able to tell when the 4 P.D. base-out prism is introduced. This is because the test will cause the image on the retina to *move* only a small amount (only four "diopters' worth"). The image will still fall within the suppression scotoma. The image over the suppressing eye can be caused to move in one of two ways. Placing the prism directly over the suppressing eye will cause the image to move, but since the movement is not perceived, the eye will make *no* attempt to re-fixate. Neither eye will move when the prism is placed over the suppressing eye. The second way to move the image is by placing the prism over the habitually fixing eye, which causes the retinal image to move off the fovea. The fixing eye ordinarily will move inward (in the direction of the prism's apex) so as to pick up fixation. The other eye will move (in a version) with the fixing eye in accordance with Hering's law of simultaneous innervation. The retinal image continues to move within the suppression region, however, so *no* convergence movement is made. This would be a positive four diopter BO test because it indicates suppression.

Practice with a *six* diopter prism until the movements are more easily observed. The four diopter BO test is extremely useful when determining the presence or absence of fusion in young patients, dishonest patients, or patients who have monofixational syndrome.

Amblyopia

Amblyopia is defined as decreased vision in one eye (though it sometimes occurs both eyes) that cannot be attributed to a specific organic problem and that cannot be improved with corrective lenses. It is like suppression that persists under *monocular* conditions.

Amblyopia, like the other adaptations to strabismus, is more likely to occur and will be more severe depending on how early the adaptation started and how long it has gone untreated.

Types of Amblyopia

Amblyopia may be categorized in several ways. It may be considered *functional*, which implies that it is treatable, or it may actually be *organic*. Organic amblyopia is considered an amblyopia because the actual organic problem is undetectable; the eye seems normal except for decreased vision, which does not respond to appropriate treatment.

Types of Amblyopia:
1. Organic
2. Functional: strabismic anisometropic, refractive, meridional, amblyopia of disuse

Von Noorden classifies functional amblyopia by their different causes. (von Noorden, 1985). Nonalternating strabismus is both a *cause* of amblyopia and a sign that amblyopia *exists*. So sometimes it is difficult to determine which came first. Nonalternating strabismus causes **strabismic amblyopia.** This type results when the nonpreferred eye is suppressed and does not develop visually. Strabismic amblyopia is fairly common in accommodative ETs, where there is frequently a preferred eye. It is treated by patching the preferred eye until absolutely equal vision is achieved.

Anisometropic amblyopia occurs when each eye has a different uncorrected refractive error. As a result, only one eye may see clearly at a time. A child with myopia in one eye often uses that eye for near fixation. The *least* myopic, plano, or hyperopic eye is used for distance fixation. Amblyopia often does *not* occur routinely in these children; they do lack fusion, however, because *both* eyes cannot see clear images at the same time. Most often, an anisometropic child with no myopia never uses the *most* hyperopic eye. Therefore, amblyopia develops. Whether looking at distance or near, the child will choose the eye requiring *less* accom-

EXAMPLE of balancing a refraction: Cycloplegic refraction: OD + 1.00 sphere
                OS + 6.50 sphere
OS will typically be amblyopic if a balanced correction is **not** given. The **difference** between the two eyes (5.50) must be maintained. A "balanced" Rx could be: OD plano, OS +5.50 sphere so as the right eye must accommodate one diopter, the left eye will also accommodate one diopter, which will allow it to see clearly.

Brands of Patches used in Amblyopia Treatment:
1. Opticlude
2. Coverlet
3. Elastoplast

modation (the least hyperopic eye) to fixate. Fully *balanced* correction of the dioptric refractive difference between the two eyes is prescribed. Occlusion may still be necessary.

Refractive amblyopia is characterized by poor visual development in *both* eyes (possibly asymmetrical), and it results from uncorrected high refractive errors. High hyperopes and extreme myopes may develop bilateral refractive amblyopia. Children with uncorrected high astigmatism may develop **meridional amblyopia.** In this case, part of the retina receives a clear image while a band of retina 90° away receives a blurred image due to the uncorrected astigmatism. As a result, a "band" of amblyopia develops in the part of the retina correspondent to the cylinder axis. Treatment requires fully balanced correction of the refractive error, and occlusion if necessary.

Amblyopia of disuse (amblyopia ex anopsia) results from disuse of the eye once the initial cause of disuse is removed. Examples are congenital ptosis, where the lid consistently covers the visual axis; congenital cataract; retinal hemorrhage; anterior chamber hyphema; etc. These maladies cause decreased vision, but once they are removed or resolved, amblyopia may persist because of the length of time the eye spent in disuse. Treatment is patching the better eye with full optical correction as needed.

Diagnosis of Amblyopia

The diagnosis of amblyopia is made by detection of a best-corrected acuity difference between the eyes that, technically, is two lines or more different. While wearing the proper correction, one eye sees poorer by two lines of acuity (i.e., OD 20/20, OS 20/30 or OD 20/15, OS 20/25). For practical purposes, particularly after amblyopia treatment has started, *any* acuity difference is considered amblyopia. Treatment is not discontinued or decreased until there is no further improvement after three to six months of occlusion.

It is imperative when testing to remember the characteristics of an amblyopic eye. It sees well under **mesopic** conditions—between dark (**scotopic**), and light (**photopic**). It also has difficulty zeroing in on one letter in a presentation of several (ignoring the other letters enough to see it).

The amblyopic eye exhibits **separation difficulty,** or the **crowding phenomenon.** For this reason, acuity is often poorer when tested by a full-line presentation of characters, rather than when tested with singly presented characters. In fact, the eye can be so much better when tested by single optotypes that amblyopia is not apparent. Any child suspected of amblyopia (therefore, any child) should have acuity tested using FULL-line presentation and *not* single Allen cards, Sheridan Gardner cards, or E cards. When testing some young children, the only test that can be used is the Allen pictures. Allen pictures are available in full-line presentation on a projected slide manufactured by American Optical Co. If this slide is not available, present the pictures in a line presentation by holding three cards together (see Figure 4.3). Always note in the chart exactly which test is used with a young child. Occasionally a two-year-old can do Snellen, but a normal six-year-old can do only E-line.

All **Allen card** pictures are made with the same size character and each subtends 5 minutes of arc at 30 feet. Therefore, the denominator is always 30, and the numerator of the acuity fraction is always the maximum number of feet away that the child could accurately identify the pictures. This value is mathematically calculated to a standard 20/X acuity fraction. For instance, $^5/_{30}$ indicates that the child accurately indentified the pictures at five feet, but not farther; $^5/_{30} = {}^{20}/_{120}$.

The **projected Allen line chart** and near vision card both have pictures of decreasing size, the smallest of which is a $^{20}/_{30}$ symbol (when tested at 20 feet for distance, or 14 inches for near).

When a patient seems too young to do Allen pictures, send home a xeroxed copy of the pictures for mom or teachers to help the child practice identifying. Before testing vision, establish that the patient is able to perform the test prior to actually doing it. Encourage the child to identify the symbols or the colors before the patch or glasses go on, the lights go out, and he finds himself at center stage.

The **HOTV** and **STYCAR** matching tests (Snellen Test for Young Children And Retardates) are two vision tests where the patient does not have to know his letters or need to talk. They use Snellen-like letters in a full-line presentation on a

Allen picture card vision equivalents:

AT, RECORDED, EQUIVALENT:
1 foot, $^1/_{30}$, $^{10}/_{600}$
2 feet, $^2/_{30}$, $^{20}/_{300}$
5 feet, $^5/_{30}$, $^{20}/_{120}$
10 feet, $^{10}/_{30}$, $^{20}/_{60}$
15 feet, $^{15}/_{30}$, $^{20}/_{40}$
20 feet, $^{20}/_{30}$, $^{20}/_{30}$
25 feet, $^{25}/_{30}$, $^{20}/_{24}$
30 feet, $^{30}/_{30}$, $^{20}/_{20}$

Figure 4.3 Three Allen picture cards held together to elicit crowding phenomenon.

STYCAR vision equivalents:
(20 feet = 6 meters)

$^6/_{60}$ EQUALS:	$^{20}/_{200}$
$^6/_{36}$	$^{20}/_{120}$
$^6/_{24}$	$^{20}/_{80}$
$^6/_{18}$	$^{20}/_{60}$
$^6/_{12}$	$^{20}/_{40}$
$^6/_9$	$^{20}/_{30}$
$^6/_6$	$^{20}/_{20}$

chart. The patient holds a card with the possible choices of letters and is instructed to match the letter on the big chart with one exactly like it on his card (Figure 4.4). The test requires two examiners; one to hold the test chart at 10 (3 meters) or 20 feet (6 meters) and point to the letters to be identified, and another to stand behind the child and silently signal if the child correctly matches the letter. The HOTV test lines are marked like a Snellen chart corrected for the 10 foot testing distance. The STYCAR testing distance is 20 feet (6 meters), so acuity is always $^6/$ the number on the last line seen.

The **Sheridan Gardner** is a matching test but has the disadvantage that letters are presented singly, and they are in ring binder form so they cannot be held together as the single Allen cards may be.

E-game may also be in card form, where the cards each display a different size E. These cards should also be held together to elicit the crowding phenomenon. Slides of full-line presentation are also available. Instruct the patient to refer to the E as "the legs of a table" and to point which way the legs go. Practice first with a large "E" before any occlu-

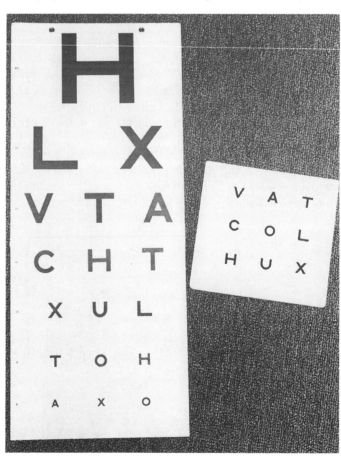

Figure 4.4 STYCAR chart (left) to be used at 6 meters (20 feet). Small card (right) to be held by child to be used to match letters seen.

sion for monocular testing, and instruct the parents to practice with the child at home, if necessary.

Objective vision tests must be resorted to for the nonverbal child or infant. These tests are gross and mainly identify a *difference* in acuity between eyes, without establishing an actual acuity level. The exceptions to this are the **visual evoked potential** (VEP) and the **preferential looking technique** (PL).

VEP measures cortical activity of the occipital cortex when the patient is presented with a patterned stimulus. The PL estimates acuity by determining when a patient looks at a patterned stimulus, which is paired with a blank stimulus that is matched in luminance (Sokol, 1986).

The **four diopter prism base-out test** is another objective test that helps determine eye preference. It requires fairly steady fixation and, therefore, is often difficult to perform on a young child (see Chapter 4).

The **binocular fixation pattern** (BFP) is a method of rating eye preference in the presence of an obvious tropia. A child with extremely unequal vision will show great preference for the good eye. A child with nearly equal vision will have only mild preference for one eye (Zipf, 1975.) (See Table 4.1.)

Prisms may be used in a variety of ways to *induce* a tropia, thus allowing the binocular fixation pattern to be assessed.

The **25 diopter base-IN test** (Cassin, 1982) induces a large ET that cannot be overcome by most children, and results in diplopia. A 25 diopter base-in prism is introduced over one eye and the child's eye preference is noted. The prism is then placed over the other eye and eye preference is noted. A child with equal vision will ordinarily use the eye *without* the prism to fixate regardless of which eye is viewing through the prism. If a child shows a preference for one eye to fixate through the prism, the non-preferred eye is considered amblyopic.

Ten-diopter vertical prisms are also used to produce a tropia with diplopia so that eye-fixation preference can be assessed (Wright, Walonker and Edelman, 1981).

Rating the monocular fixation pattern as *central, steady, and maintained* provides limited information. An eye with extremely poor visual acuity may also have central, steady, and maintained fixation.

Prism Tests to Induce Tropias:
1. 25 Diopter Base IN Test
2. 10 Diopter Vertical Prism Test

Table 4.1 Binocular Fixation Pattern

Number 5	Spontaneously alternates
4	Holds fixation through blink
3	Holds fixation until blink
2	Holds fixation 1–2 seconds but switches before blink
1	Immediately switches fixation, strong preference

Fixation Patterns

Paradoxical fixation: Or-
dinarily, an esotropic patient
would use an eccentric point
in his **nasal** retina and an exotropic patient would use a
point in his **temporal** retina.
The fixation is considered to
be **paradoxical** when an ET
uses temporal retina or
when an XT uses nasal
retina.

Some young children develop **eccentric fixation,** which
means they use a point other than their fovea to fixate
monocularly. Because the acuity potential of the peripheral
retina is lower than that of the fovea, a patient with eccentric fixation will always have decreased vision. These patients
"think" and feel that they are looking straight ahead; that is,
the eccentric point has assumed the role of the fovea. When
fixating, they move their retina so the eccentric point faces
the object of regard. Signals of eccentric fixation include
poor acuity AND the appearance that the eye is not fixing
centrally (see Figure 4.5). Some children may appear to fixate
eccentrically because they have dragged retinas resulting
from retinopathy of prematurity. This can be confirmed by
funduscopy. They are actually foveally fixing but, because
the fovea is not located in the usual position, it appears as
though they are eccentrically fixing.

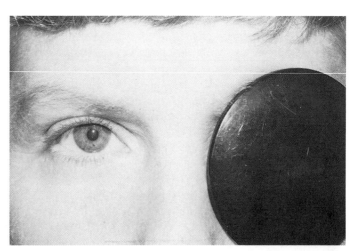

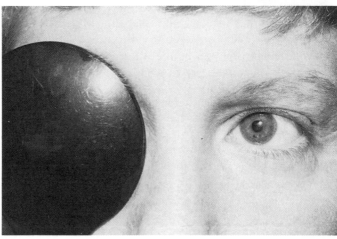

Figure 4.5 (Top) monocular
fixation with right eye is
central. (Bottom) monocular
fixation with left eye is
eccentric (not central).

A variation of eccentric fixation is *eccentric viewing*. Some patients, thinking the fovea should be used for fixation, will often use it initially when asked to look at something. Because the fovea has poor acuity, however, they shift to fixate with an eccentric point. Although this point usually yields better acuity, patients don't "feel" as though they are looking straight ahead. Patients with central macular defects who fixate with an eccentric point off of the defect could be considered to be eccentrically viewing.

A young child with eccentric fixation can often be successfully treated with ordinary occlusion. Eccentric fixation may be so in-grained in older patients that occlusion of the sound eye only encourages use of the amblyopic eye's eccentric point. In these cases, **inverse occlusion** may be used, where the amblyopic/eccentric eye is occluded in the hope it will "forget about" the eccentric point. Once central fixation is reinstituted, direct occlusion of the sound eye is done.

Treatment of Amblyopia

Amblyopia is treated by forced use of the amblyopic eye. Refractive correction is worn and the better-seeing eye is occluded. An eye patch is used directly on the face and full-time occlusion is usually instituted first. Instruct the patient's parents to put the patch on first thing in the morning and to take it off last thing at night. Tell them if the child is up for an hour before the patch is put on (or without the patch for an hour before bedtime), that each hour spent without the patch undoes the whole day. While there is no solid proof of this, it gets the message across to the parents!

Also tell the parents that it is very easy for the ophthalmologist to tell them to go home and make their two-year-old wear a patch (or glasses), but that it is understood who has to go home and do it. Be sympathetic, but tell them that wearing the patch is no different from getting the child to wear clothes, not run out in the street, and to refrain from putting fingers into electric sockets. Wearing the patch is in the same category of these "Big No's." Tell the parents that amblyopia is a curable disease, but that it must be treated while the child is still young, which means too young to make a rational decision. The parents are obligated to keep their child wearing the patch. If they don't follow through, *they* will have to live with the responsibility for their child's blind eye.

After the seeds of guilt are carefully sowed, make it clear to the parents that they can call the office for moral support.

As a general rule, a child on a full-time patching schedule should be re-evaluated at time intervals dependent on age: one week per year of life. Although dense amblyopia is rarely cured in the first patching trial, the threat of **reverse** or

Patching follow up: a three-year-old child should be checked in three weeks and a six-month-old infant in three or four days, with no one going longer than four weeks.

occlusion amblyopia is possible. When occlusion of the better-seeing eye is carried out too long (time length varies from child to child), the eye may become amblyopic.

The goal of amblyopia treatment is to achieve equal vision. Once equal vision is achieved, patching may be slowly decreased. Some form of **maintenance patching** must be continued—particularly in cases where every minute *without* the patch results in fixation by only the non-amblyopic eye—until the child reaches an age where vision will not be lost again. This age varies among children. While most visual development occurs by age 5, for many children this occurs by age 6. Some children may not reach visual maturity until age 8 and, occasionally, some show variation until the teen years. For this reason, parents should be told that patching in some form, will go on for years—perhaps until the child is a teenager.

There is some evidence that the visual systems of anisometropic amblyopes are more elastic at a later age. These patients can be successfully treated at a later age than other amblyopes. However, therapy must also be continued for a longer period.

Patching may be decreased from a full-time wearing basis in a variety of ways. The patch may come off for a few hours at a time each day. Time spent without the patch is increased as subsequent checks reveal vision is remaining equal despite decreased patching. Eventually, patching may be necessary for only an hour a day to maintain equal vision. Ask the parents when it is most convenient for them to decrease patching (i.e., removing the patch at supper time, or before nursery school). A child with an intermittent eye turn is most likely to be fusing early in the day and tropic later. Instruct the parents that you want the patch worn when the child is most likely to be tropic, or late in the day. Every other day patching works well for some children.

Eventually, **partial occlusion** may become a practical solution. This involves significantly blurring the vision of the dominant eye so that fixation is switched to the non-preferred eye. Spectacle lenses may be blurred by adding clear contact paper, scotch tape, a high plus Fresnel press-on lens, graded Bangerter film, or simply by prescribing a more plus correction than needed. A high plus contact lens could also be used in the dominant eye. All of the partial occlusion methods should *only* be used on an older, cooperative child who has already achieved equal vision by traditional patching. These methods are NEVER to be offered as an alternative to a child who gives his parents a hard time about wearing a patch. Glasses can be removed and contact lenses can be rubbed out. A child that won't wear a patch happily won't wear glasses happily either.

Children with latent nystagmus, or other forms of nystagmus, may exacerbate their nystagmus with traditional occlusion and so may benefit from partial occlusion in some cases.

Penalization is a method of using drops and glasses to optically force one eye to be used, either for distance fixation or near fixation. Again, an uncooperative child can remove glasses and, unless the refractive error is in your favor, cyclopleging the fixing eye simply does not blur it enough to force fixation with the amblyopic eye.

Finally, watch out for those children who cheat on vision tests, either intentionally or unintentionally. Occasionally, an older child will come in for a routine eye exam and much to the parents' surprise, be found to have profound amblyopia. Invariably, it is the left eye that is amblyopic in these cases. The child will confess to having known about the poor vision for years, and will go on to explain how he got by all this time. Such patients were routinely asked to read the $^{20}/_{20}$ line with the right eye. The occluder is then switched, and they were asked to read the same $^{20}/_{20}$ line with the left eye. Vision testing requires more variation than that. Ask patients to read the line backwards, or to identify what the third letter from the end is. Don't let them fake their way through the screening!

Abnormal/ Anomalous Retinal Correspondence (ARC)

Normal binocularity results in excellent stereopsis with bifoveal fusion (see Chapter 1). Normal retinal correspondence (NRC) exists when the fovea of one eye corresponds to the fovea of the other so the two images blend into one fused image.

ARC may develop as the brain attempts to develop a new binocularity to adapt to strabismus. Although ARC is not entirely "normal," it is a form of binocularity that yields gross fusion and, therefore, a sort of stability to the eye turn. Invariably, it coexists with some sort of suppression, and may or may not be accompanied by amblyopia. Like suppression and amblyopia, ARC is not "all or none"; it has levels of severity. The longer the strabismus exists, and the younger the child, the more likely that ARC will exist. It also exists more frequently in patients with smaller eye turns than in patients with larger ones, because the fusion yielded with smaller eye turns is better. Usually, ARC is not an advantageous adaptation to strabismus that is larger than thirty prism diopters.

ARC occurs when the fovea of one eye corresponds to a *nonfoveal* point in the other. Images stimulating these corresponding retinal points will be superimposed and/or fused. A form of gross binocularity results, despite a strabismus.

ARC may coexist in a patient with NRC. For instance, an X(T) patient may have NRC and bifoveal stereopsis when

Types of Retinal Correspondence:
1. Normal
2. Abnormal/Anomalous:
   Harmonious
   Unharmonious
   Paradoxical

Figure 4.6 Bagolini lenses in trial frame right lens set at 135°, left lens set at 45°.

straight (see Chapter 3), but suppression and ARC when tropic. The result is that the patient has no noticeable symptoms when tropic; the only *sign* of ARC is the XT eye position.

Another example of ARC and NRC coexisting is in a patient with incomitance. A patient with a V pattern ET may exhibit NRC and perfect fusion in up gaze, where the eyes are straight, and ARC in down gaze, where they are esotropic.

ARC varies and is not "completely present" or "completely absent" in any given patient. Therefore, "depth" of the ARC is best determined by different tests.

Diagnosis of ARC

Any subjective fusion test where fusion is reported in the presence of a tropia indicates ARC. **Bagolini lenses** are plano lenses with fine striations that cause a light to appear as a streak perpendicular to the striations (like a Maddox rod). The lenses are used in a trial frame and set so that one is axis 45°, the other is axis 135° (see Figure 4.6). A patient who reports a perfect cross is reporting fusion, and fusion in the presence of straight eyes indicates NRC (see Figure 4.7). A cover-uncover test that reveals a tropia while the patient reports fusion indicates ARC. Since Bagolini lenses are most like real seeing conditions, the presence of ARC by Bagolini lenses is frequently detected in young patients with a tropia.

Either **Worth Four Dot** (W4D) testing or a haploscopic instrument, such as the **troposcope** or **synoptophore,** may be used to test ARC (see Figure 4.8). The W4D may be difficult to interpret, because the examiner cannot be absolutely certain a tropia exists at the moment a patient reports fusion. For instance, a child with accommodative ET might *not* be accommodating on the W4D lights. The child may therefore have straight eyes during W4D retinal correspon-

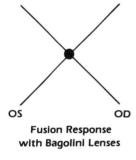

OS   OD
**Fusion Response
with Bagolini Lenses**

Figure 4.7 Patient's view of light while looking through Bagolini lenses in the presence of fusion. If patient is ortho, this fusion response indicates NRC; if a tropia exists, response indicates ARC.

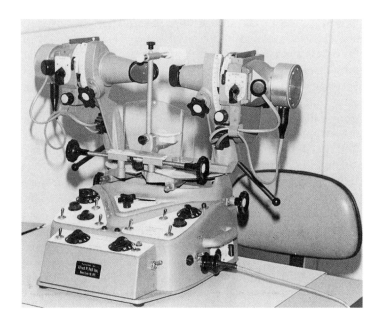

Figure 4.8 Synotophore type of haploscopic instrument.

dence testing and, have NRC. Conversely, if the patient does accommodate (and go esotropic) during testing and still report fusion, ARC exists.

It is easier to determine the presence of an eye turn during haploscopic testing. Ordinarily, the patient's objective angle is measured by cross-covering and neutralizing eye movement. This is done by moving the arms of the instrument that project the images (instead of using a correcting prism). The **objective angle** is the actual measurement of the patient's eye turn. Once the objective angle is measured, the picture targets fall on each fovea. If NRC exists, the patient will report superimpositioning of the targets. If ARC exists, the two images will fall on noncorresponding retinal points, and diplopia will be reported. The arms of the instrument can be adjusted to move the images onto corresponding retinal points, where superimpositioning occurs. The eye position where *subjective* fusion occurs is the **subjective angle.**

Although use of the haploscopic instrument is the easiest way to demonstrate **harmonious** and **unharmonious** ARC, the same can be done using the W4D or Bagolini lenses and a correcting prism. To *understand* the concept, however, it is easiest to think about what happens on the haploscopic instrument.

Basically, ARC is the brain's fusional adaptation to strabismus. Although the eyes are tropic, the brain "thinks" the eyes are fusing as though they were straight. *Harmonious* ARC exists when the brain thinks that the eyes are straight; that is, subjective fusion occurs when the retinas are stimulated as they normally are in a tropic patient. If the

To measure the patient's subjective angle during W4D testing, place correcting prism over one eye until subjective fusion is reported or until the diplopia shifts sides. If the amount of prism used was **equal** to the size of the deviation when measured by P + C (the objective angle), then NRC is present. If no prism was necessary (the patient reported fusion in the presence of a tropia), then harmonious ARC is present. If the prism used was **less than** the size of the deviation when measured by P + C, but **more than** zero, then **unharmonious ARC** is present.

After Image Testing Response

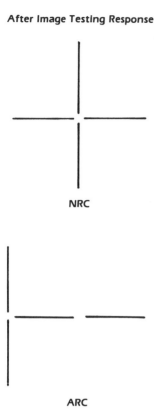

NRC

ARC

Figure 4.9 Patient view of after image testing.

Maddox rod measurements will be affected and are not valid in the presence of ARC.

patient is 30 P.D. ET, and reports *fusion* when the fovea of one eye is stimulated and a point "thirty diopters" nasal to the other fovea is stimulated, harmonious ARC exists. That patient would also report fusion to W4D testing and Bagolini lenses, and when the arms of the haploscopic instrument are set at *zero*. When a patient has an objective angle that represents a tropia and a subjective angle at zero, harmonious ARC is present.

If the subject angle is anything between zero and the objective angle, the patient has *unharmonious* ARC. It is as though the patient's retinal correspondence is somewhere between NRC (subjective angle = objective angle) and harmonious ARC (subjective angle = zero/ortho).

The presence of unharmonious ARC should be documented because it is thought to represent an in-between stage for the patient shifting from NRC to full ARC. Any type of ARC is important to document because it is thought to affect surgical outcome. A patient with ARC whose eyes are suddenly straightened by surgery may be diplopic postoperatively. With straight eyes, the object of regard falls on or near the fovea in each eye. The foveas, however, are not corresponding retinal points and when they are stimulated, diplopia occurs. In most young patients, diplopia of this nature disappears within a few days. In older patients (fifteen years-old or so), there is a small chance that diplopia may persist, either in the form of **intractable diplopia** (which goes away only with occlusion), or in the eyes drifting back to the original eye position where the ARC angle exists. The depth of the ARC also helps predict which patients will be most bothered by postoperative diplopia.

Afterimage testing for ARC employs the use of conditions most unlike normal seeing conditions, so a patient who has ARC with afterimage testing has a well-established ARC pattern. If this is found, particularly in an older patient, the chances of postoperative diplopia are greater.

Unlike the previously mentioned ARC tests, the afterimage test is NOT a diplopia test. Diplopia tests involve one object of regard (the W4D flashlight or the penlight used with Bagolini lenses or with Maddox rod testing). The diplopia test then dissociates the eyes so diplopia may be appreciated and so each eye is aware of an image.

The afterimage test is a "fovea to fovea" test where there are two actual images, one presented to each eye. A bright light filament is shown to each eye so as to leave an afterimage effect on the eyes, open or closed. A small, central area of the filament is covered, and the patient is instructed to fixate there. (This is done both to protect the fovea from the bright light and afterimage, and to control fixation.) The usually fixing eye only is exposed to the filament horizontally-held. The filament is then turned and presented vertically to the habitually deviating eye only. Each fovea now has is own image. If the patient has NRC, a plus sign will be perceived, regardless of eye position. If the foveas do not correspond to each other, no intersection of the afterimages will occur (see Figure 4.9). Have the patient draw what he sees and blink a lot to make it easier for the afterimage to be seen. Although the afterimage can be seen in a lighted room, it may be more obvious in the dark. The diagnostic value of comparing the response in the light versus in the dark is questionable.

Documenting ARC with the different tests may help predict which older patients will have postoperative diplopia. While all of these tests may provide some input, I find it easiest to hold up a correcting prism on an older patient with ARC when contemplating strabismus surgery. A patient with harmonious ARC who measures 30 P.D. ET may be diplopic with 30 P.D. base-out correcting prism. Although this puts the image on both foveas, the fovea of the deviating eye does not correspond to the fovea of the other eye. In harmonious ARC, a point thirty diopters *nasal* in the deviated eye corresponds to the fixing fovea. The fovea of the deviating eye is then considered to be relatively "temporal" to the nasal point that corresponds to the fixing eye's fovea. Now that the image falls on the deviating eye's fovea (which is considered "temporal"), the patient will report *crossed* diplopia (the objective angle). (See Figure 4.10.) The use of any prism larger than thirty will result in crossed diplopia with the images getting farther and farther apart. If prisms smaller than thirty base-out are used, the images will get closer together until diplopia either disappears or "crosses over" to uncrossed diplopia. The point at which the diplopia crosses over is the patient's subjective angle. If the angle is zero, the patient has harmonious ARC. If it is in between zero and thirty, the patient has unharmonious ARC and that part of the eye turn could be surgically corrected without a large risk of postoperative diplopia. If the patient crosses over with twenty base-out, 20 P.D. of the original eye turn could be safely corrected. This would leave the patient with a 10 P.D. ET, which is cosmetically acceptable when compared to the 30 P.D. ET that existed before surgery.

Paradoxical ARC describes an ET patient whose subjective fusional angle occurs when the *temporal* retina of the deviated eye is stimulated. (The fovea of the fixing eye corresponds to a temporal point in the esotropic eye.) Ordinarily, a *nasal* point of the deviating eye would correspond with the fovea of the other eye in an ET patient. Paradoxical ARC usually results from consecutive deviations where the patient was exotropic to begin with, had surgery,

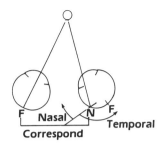

Figure 4.10 Harmonious ARC: RET causes the object of regard to fall on a nasal point (N) of the right eye. (N) corresponds with the fovea of the left eye. The right eye now considers all points nasal of (N) to be "nasal" and all points temporal of (N) to be "temporal" (even though the retina between (N) and the fovea of the right eye is really **nasal** retina).

then became esotropic while maintaining the original ARC angle. Paradoxical ARC may also occur in an XT patient whose subjective fusional angle occurs when *nasal* retina is stimulated.

Summary

Both a patient's history and fusional response in the presence of a tropia help signal whether that patient has ARC. The history would indicate a long-standing tropia that was present from a young age. The tropia may be intermittent or constant, but is usually less than 30 P.D. The patient would not have bifoveal stereopsis, (see Chapter 3) and would report fusion on tests like the W4D. The tricky part comes when an esotropic patient reports uncrossed diplopia, which you would expect, but the diplopia is not appropriately separated. A patient with a 30 P.D. ET should have uncrossed diplopia, but it should be "thirty prism diopters worth." If the uncrossed diplopia is close together, unharmonious ARC probably exists. Prisms should be used in conjunction with the W4D lights to determine exactly where the patient's subjective angle is located.

The implications of ARC should be considered mainly when surgical correction of the eye turn is planned. The technician's awareness of ARC is also imperative when doing subjective tests because it will influence the measurements.

Treatment

In the past, ARC was routinely "treated" preoperatively on most patients. The goal was to re-establish NRC so the patient would not be diplopic and would have an increased chance of fusion postoperatively. Surgery is generally done at much younger ages now—either before significant ARC develops, or before the ARC angle is well established. For this reason, ARC is not actively treated. Alternate occlusion eliminates the patient's need for suppression and for ARC, as those are both adaptations to *binocular* strabismus. Some people therefore alternately occlude children preoperatively both to ensure equal vision and good ocular rotations, and because it may help the ARC.

References

Cassin, B. Alternate fixation in the non-strabismic child. *American Orthoptic Journal* 32:111–116, 1982.
Sokol, S. Alternatives to Snellen acuity. *American Orthoptic Journal* 36:5–10, 1986.

von Noorden, G.K. *Burian — von Noorden's Binocular Vision and Ocular Motility.* 3rd ed. St. Louis, C.V. Mosby 1985, p 210.

Wright, K. Walonker, E. Edelman, P. 10 diopter fixation test for amblyopia. *Archives of Ophthalmology* 99:1242–1246, 1981.

Zipf, R. Binocular fixation pattern. *Archives of Ophthalmology* 401–405, 1976.

CHAPTER 5

History Taking

Goals of History Taking

Questions about a patient's history are asked to formulate a tentative diagnosis. The exam that follows involves various tests that will eventually lead to a final diagnosis. These tests are dictated by the tentative diagnosis, which was a direct result of the history. In other words, the history of the patient sets up the exam. Your goal is to be a good detective.

Treat the questioning as if you are playing a game of twenty questions, where one question leads to another until the tentative diagnosis is formed. Patients are frequently unable to put their current symptoms together with their history. For instance, they would never think to mention that they are being treated for arthritis; they would never guess that a drug such as plaquenil could affect their eyes. It is the history taker that puts two and two together for the patient—but only if that history taker asks appropriate questions.

Approach each question knowing what the potential answers can be and what each of those answers will imply. When you innocently ask, "Does your lid droop worsen with fatigue?," you are *really* asking, "Could you have myasthenia gravis?"

The tables of this chapter are to be used as a basis for your questions. Eventually you will keep tables of your own in your head. Recheck a patient's chart after the examination and diagnosis. Compare your history and tentative diagnosis of the patient with the final diagnosis. Then decide if you really asked appropriate questions.

Typical Questions of Any Ophthalmologic Patient

Primarily, you are searching for why the patient has an appointment. You need to ask about current **main eye complaints** and then about any other **secondary eye complaints** the patient may have. Ask about any **precipitating factors:** what could possibly be causing the eye condition. Then you can ask about the patient's **past ophthalmic history,** including treatments or surgery. (Guidelines for general ophthalmologic history questions are thoroughly covered in the Ophthalmic Technical Skill Series - Basic Skills).

Once the eye history is completed, you need to inquire about the patient's general history. Ask about the patient's *health* with specific questions. Merely asking, "Are you in good health?" is not adequate because anyone with controlled diabetes will honestly answer "yes." Inquire about any *medications* and any known or suspected *allergies* (Mausolf, 1975).

Typical Questions of Children

Children require an additional set of questions. Every child's history should include questions about **birth history, systemic disease** or **congenital syndromes** and **developmental milestones.** Birth history questions are covered in Table 5.1, systemic disease or congenital syndromes questions are covered in Table 5.2 (Harley, 1975), and developmental milestones questions and normal developmental milestones are covered in Table 5.3 (Kempe, Silver, and O'Brien, 1984).

Typical Questions of Any Patient With Strabismus

An eye turn, or **strabismus,** is often a patient's "main complaint." There are five areas to be covered and each requires astute questioning, deciphering of the answers, and physical observation. Always watch the patient with the supposed ocular motility problem while discussing the problem. If it is a child, do the questioning and observing somewhere besides the exam room or exam chair. Children

Typical questions of any ophthalmologic patient—
Main complaint: What brought the patient in to the office today?
Secondary complaint: Are there any other eye problems that have been bothering them?
Precipitating factors: Has the patient noticed anything that makes the eye problem worse (i.e., fatigue, certain medications, foods)?
Past ophthalmic history: Have they had any previous eye problems? What kind of eye treatments have they had previously (i.e., glasses, patching, surgery, medications)?

Typical concerns of any patient with strabismus
1. Cosmesis
2. Vision
3. Strabismus
4. Diplopia
5. Family History

Table 5.1 Birth History

- Was the pregnancy normal?
    Any complications; toxemia, bleeding, illness, rubella?
    Was there any alcohol, tobacco, or drug use?
- Was the delivery normal? Type of delivery? Length of labor?
    Any complications: fetal distress or use of forceps?
    If forceps were used, were there any marks on the
    head or over the eyes?
- Was the baby premature? < 2500 gm. Normal gestation is 34–40
    weeks.
- What was the 1 minute APGAR score? 5 minute APGAR score?
    Perfect APGAR is 10. (APGAR *score* evaluates: heart rate,
    respiratory effort, reflex irritability, muscle tone, and
    color)
- Did the baby have oxygen and how much for how long?
- Were the retinas checked while in the hospital or since then?
- Was there any other treatment at or shortly after birth?
- Was the baby healthy enough to go home in a few days?

Table 5.2 Systemic Disease or Congenital Syndromes

Does the child have any systemic disease or congenital syndrome?

	May cause:
Down syndrome	strabismus, refractive errors, amblyopia
albinism	nystagmus, blindness, photophobia
hydrocephalus	setting sun phenomenon, strabismus
spina bifida	strabismus
meningomyelocele	strabismus
juvenile diabetes	retinopathy, "sugar" cataract
juvenile rheumatoid arthritis	Brown's superior oblique tendon sheath syndrome
neurofibromatosis	optic nerve gliomas
Apert syndrome	bilateral SO palsies, XT, hypertelorism, astigmatism
Crouzon syndrome	XT, shallow orbits
Marfan syndrome	dislocated lenses, myopia
Grave's disease	restrictive hypertropias, decreased elevation, proptosis, exophthalmos
Sturge-Weber syndrome	glaucoma, amblyopia

Table 5.3 Developmental Milestones

sucking reflex: at birth
first raised head: by 2 months
rolling over: 2–5 months
sitting up: 5–8 months
pulling up to stand: 6–10 months
standing: 9–13 months
walking: 11–14 months
babbling words: 12–20 months
short sentences: 14–24 months

only sit still in the chair and cooperate for a limited number of minutes. Don't waste precious minutes asking history questions, which could be done *before* the child is in the exam chair.

Cosmesis

Determine the presence of a cosmetic problem, either an eye turn or an unusual head position. Also find out who notices the problem. Grandmothers are usually very good at first noticing strabismus, but an old picture (driver's license or baby picture) may tell you more. Be wary that pictures can be deceiving because of poor quality or head positions forced by the photographer. A picture may also have been reversed, where the right eye is really the left. Old pictures may help document the age at which the strabismus or head position started. Ask if the child squints. This could be an attempt to **pinhole,** which would improve vision in a patient with an uncorrected refractive error. A child who always closes the same eye, or who consistently sneezes and coughs when in bright sunlight may be exhibiting the typical symptoms of children with exodeviations. Determine if the child has an unusual head position, and document what it is. Ask when the head tilt occurs most frequently. (It may only be when watching TV or reading.) Has anyone noticed any rhythmic oscillations (**nystagmus**) of the eyes? Are the eyes moving when they shouldn't be? Is there a lid droop (**Ptosis**)? Finally, what is the general appearance of the patient? Is the strabismus obvious to you?

> A head position that is due to an eye disorder is termed **ocular torticollis** and disappears when one eye is occluded. A head position that persists with occlusion is termed **congenital torticollis**.

Vision

Ask the parents if they think the child sees well. How does the child get around in unknown surroundings? Watch the child maneuver in the office while you are asking these questions. No parent wants to admit that their child may not see well, so decide on your own if they are accurately describing their child's visual behavior. Has the child ever worn glasses? Were they used by the child? Be specific: don't just ask if the child *wore* them, ask if he looked *through* them. Did he use a bifocal appropriately, or look through the distance portion? Ask if the child only uses one eye because the continual use of only one eye may make the other eye develop poorly and have decreased vision [**amblyopia**]. Does the child have an eye turn where it's always the same eye that turns? Ask if an eye patch was ever worn, on which eye, and for how long. Was nystagmus ever present in either one or both eyes.

Strabismus

Are one or both eyes deviating from the normal position? Does the eye turn *inward* or *outward*, *upward* or *downward*? Does the eye seem to move freely in all directions despite being turned, or are there some positions the eye will not move into at all? Is the eye turn constant or intermittent? If intermittent, when does it happen? Is it present only at near fixation, or when the child looks into a particular field of gaze? At what age did this start? Is it associated with any other factors, such as fatigue, illness, anger, only with glasses on, only with glasses off, or with trauma, either physical or emotional. Finally, ask the parents or patient what *they* think makes it worse.

Diplopia

Double vision, or **diplopia** is a new and unusual experience for most patients. They simply do not know how to accurately describe it. The history taker must ask very specific

Types of diplopia (double
vision)
1. Monocular
2. Binocular
   Physiologic: crossed or
   uncrossed
   Pathologic: horizontal,
   vertical, torsional, (tilted)

questions to get the patient to describe diplopia in a useful way.

Because **monocular diplopia** is invariably caused by an ocular media obstruction and NOT an ocular motility problem, you must determine if the patient has monocular or **binocular diplopia.** Monocular diplopia would *persist* with one eye closed and binocular diplopia would *disappear* with one eye closed. If it is truly a binocular phenomenon, ask if it came on suddenly or gradually, either painlessly or following trauma. A true sudden onset of diplopia often implies a cranial nerve palsy, which causes a specific extraocular muscle (EOM) to be underactive. The trauma could have been subtle. Ask specifically if the patient may have hit his head on a cupboard or been in a minor automobile accident. Ask if the patient has thyroid or myasthenia gravis if you missed that question in the initial questioning.

Cranial nerve palsies usually cause the deviation to be *incomitant*; the deviation varies depending on where the patient looks. So, while the strabismus could be called "intermittent," the intermittency is caused by the patient's direction of gaze. As the eye turn varies, the diplopia changes.

Another type of intermittent diplopia can occur when normal, binocular use of the eyes together is broken down. An intermittent strabismus breaks down into its tropic state during a patient's occasional loss of fusion. The breakdown of fusion is often caused by fatigue and associated eye strain. Carefully question the patient about the intermittency of their diplopia—is it due to eye position variation or fatigue?

Ask what the patient does to make the diplopia go away. Some patients have to close one eye while others will only need to "re-adjust" or "re-focus," or to blink a few times. Try to determine if the patient is searching for a new head position to get rid of the diplopia.

Finally, ask your patients to describe the diplopia itself, without putting the words in their mouth. Ask only if it is always side to side (horizontal diplopia), or up and down (vertical diplopia). Virtually any vertical deviation will give the patient a **combined** horizontal and vertical diplopia. If the patient claims that it is strictly vertical, you should question that answer. The patient may also complain of a torsional component where one image appears tilted. This would indicate that a **cyclovertical muscle** palsy exists (see Chapter 3).

When questioning about diplopia, keep in mind that if it were caused by an accident, that the patient may be in litigation for that accident and, therefore, the patient may not be absolutely truthful in answering. Many patients initially deny that there was any trauma but later ask questions about whether their double vision *could* have been caused by being in an automobile accident, or by falling off of their neighbor's poorly lit porch. Be cautious if you suspect that the patient is in litigation. Also, keep in mind that you will need to use **objective testing methods,** (where the examiner determines the results) as opposed to **subjective testing methods,** (where the patient interprets the results). Use objective prism and cover measurements instead of subjective Maddox rod measurements to assess such a patient.

Family History

As with any eye problem, ask if the patient has any blood relatives with a similar problem. They have probably forgotten about their cousins in Montana who have had strabismus surgery, or that Uncle Ernie had a congenital cataract. So ask the parents specifically about any brothers, sisters, aunts, uncles, cousins, grandparents, or themselves — who may have had similar problems. If there is a relative with a similar problem, inquire about the treatment and, most importantly, if the treatment had been successful. The patient and/or parents may already have a preconceived notion about what treatment will and will not work, or how the prognosis will affect them (Scott, Lennarson, D'Agostino, 1983).

Continually alter your questioning as the tentative diagnoses are eliminated and the final diagnosis narrowed down. Both the history taking the examination require appropriate flexibility.

References

Harley, R.D. *Pediatric Ophthalmology*. Philadelphia: W.B. Saunders, 1975, p 36.

Herrin, M.P. *Ophthalmic Technical Skills Series - Basic Skills*. Thorofare, Charles B. Slack, Inc., 1988.

Kemp, C.H. Silver, H.K. O'Brien, D. *Current Pediatric Diagnosis and Treatment*. 8th ed. Los Altos: Lange Medical Publications, 1984, p 1.

Mausolf, F.A. *The Eye and Systemic Disease* St. Louis: C.V. Mosby, 1975, p 8.

Scott, W.E. Lennarson, L. D'Agostino, D. *Orthoptic and Ocular Examination Techniques*. Baltimore: Williams & Wilkins, 1983, p 120.

CHAPTER 6

A Systematic Approach to Strabismus

Introduction

Initial baseline strabismus measurements act as a starting point and provide adequate information for making a diagnosis and planning treatment. Subsequent exams and measurements are used to look for progression or regression of the condition and to assess the effectiveness of treatment. If the treatment is found to be ineffective, it can be changed. Once the problem is corrected, occasional follow-up exams watch for recurrence, or new (related or unrelated) problems.

First determine what it is you are interested in, then organize your exam so subsequent tests will not be compromised. If the patient is a young child, be certain to get the most important information *first*. The part of a motility exam where the measurements are actually taken and assessed requires a calculated approach by the examiner. As with any exam, some tests must be done before others so as not to contaminate results. Examples of this systematic approach in general ophthalmology include taking a vision test before an applanation tonometry, or testing corneal sensation before anesthetising the cornea. In ocular motility testing, some general rules apply so that one test done too early does not alter the rest of the exam.

The other major complicating factor is the nature of the beast being examined: the child. Ideally, the entire exam should be completed in one visit. If a child is destined to mentally decompensate (go berserk) midway through the exam, it is important to have already obtained the most crucial information. If the only test completed on Scotty is a color vision test and he won't let any others be performed, color vision had better be critical to his exam. Otherwise, both the parents (and you) will feel as though the exam was a waste. Approach each individual part of the motility exam of

Learn which tests are time consuming, or frightening to a young child. **Learn** which tests must be done before others so as not to contaminate results. **Learn** to be familiar with difficult tests before a child and parent are counting on you to be "swift and painless."

a child as though it were going to be the last piece of information obtained from that child on that day. Keep asking yourself, "what test is next in importance?"

Examining Any Ocular Motility Patient

Fusion

Determining the presence of fusion is the primary goal when testing many patients. Fusion can be disrupted by certain "obstacles" that should *not* be introduced before the patient's fusional status is established. These disruptions include the removal of glasses to check power (accommodative ET will break into tropia and may not immediately be able to refuse); checking visual acuity (covering one eye for a prolonged period may break down deviation), checking versions (the bright light may dissociate a patient); or any dissociating factor, such as a red filter, Maddox rod, or Worth four dot glasses. Any of these may dissociate patients, causing them to break into a tropia. Subsequently, fusion or fusional amplitudes cannot be measured in the exam.

> In the course of a regular exam, anything that may disrupt fusion should not be done **before** the fusion is tested.

Fusion tests are either subjective—requiring a patient response—or objective—requiring the examiner to watch for signs of fusion. Subjective tests are generally the most accurate and preferable.

The least dissociating way of assessing subjective fusion is with one of various stereotests. The eyes are slightly dissociated when measuring stereopsis, so the patient is most likely to remain fusing with this method. The various stereotests include directions for grading the stereopsis tested. For Titmus stereotesting, which utilizes horizontal disparity and polarized glasses, the patients are asked to identify the circle that appears closest to them in each test group. A child may be asked to "push the circle down" with a finger. Other random dot test require identifying shapes.

> Stereotesting is the **least** dissociating fusion test available.

Stereopsis is always recorded in seconds of arc and is measured by determining the finest (most subtle) stereopsis accurately detected. Bifoveal stereopsis is stereopsis that is better than 67 seconds of arc. Be wary of good guessers and cheaters. Remember that a stereotest turned 90° has no stereoscopic affect.

The **random dot E** (RDE) and other random dot stereotests may also be used for testing near fusion (Tillson, 1985) (see Figure 6.1). The random dot tests have the advantage of producing virtually no **monocular clues** to depth perception. The child either sees the figure, or does not.

The **AO distance vectograph slide** dissociates the eyes with polaroid glasses and uses horizontal disparity like the

> **Random dot E** is a stereotest that is useful for young children who may not understand other stereotests. The child is given a choice between two cards. If gross stereopsis is present, the child will see a large "E" on one card and nothing on the other card. The child merely picks the card with the shape.

Figure 6.1 Random dot E consists of **three** cards to be used with polarized glasses. Front and back of each card is shown. Top row of cards shows test side of each card. Bottom row of cards shows reverse side of each card. Model card test side can be seen without stereopsis.

near Wirt stereotest does. Besides measuring stereopsis at distance, it can be used as a simple test of suppression. A 20/40-size line of letters is arranged so two letters are seen by either eye, two other letters are seen *only* by the right eye and two letters (six total) *only* by the left eye (see Figure 6.2). Central suppression of either eye is easily detected when the patient does not see two of the letters. The suppressed eye is determined by which letters are neglected when the line is read.

A simple test that requires no special equipment is the "pencil point to pencil point" test (see Figure 6.3). One pencil is held stationary and another is held several inches away. The patient is then instructed to touch the point of the stationary pencil with the other pencil. The patient must do this fairly quickly for an accurate measure. You can demonstrate this to yourself (providing you have good stereopsis) by trying it first with both eyes open, then with one eye closed (which removes stereopsis). While the seconds of arc cannot be recorded, the test may be helpful in a young child or other patient unable to perform the standard stereotests.

A patient with excellent bifoveal fusion cannot be constantly tropic. After stereotesting, fusion by **Worth four dot**

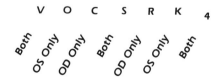

Figure 6.2 20/40 line of AO Vectograph slide.

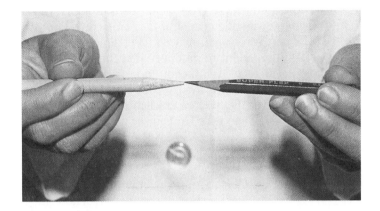

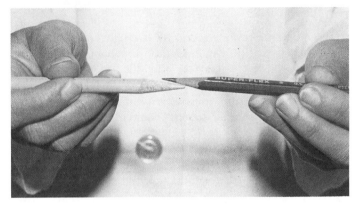

Figure 6.3 Pencil point to pencil point test for gross stereopsis. (Top) patient with stereopsis is able to touch points. (Bottom) patient without stereopsis cannot touch points.

(W4D) may be done, followed by the cover-uncover test to determine the presence of a phoria or intermittent tropia. The W4D test requires two things: your ability to interpret what the child is saying, and the child's ability to describe what he is seeing. This is a perfect example of why you must practice these tests on older, cooperative people before you try them on a younger or uncooperative person.

Before you even put the glasses on your patients, establish that they can distinguish two colors. It really doesn't matter if they call red "purple" or if green is "yellow," as long as they are consistent and you know what they are talking about. Then try to establish if they can count, at least up to three. If you start them off by saying: "one . . . two . . . " and they finish by saying "five," it really doesn't matter, as long as you interpret their answers appropriately. If a child can't count, W4D testing at distance is nearly impossible. At near you can have the child touch the lights to identify them, or use the Child's Worth lights, which have pictures instead of circles.

Patient responses may lead to the following conclusions: four lights represent fusion; two red lights represent suppression of the "green eye" (the eye with the green lens over it); three green lights represent suppression of the "red eye"; five

W4D Diplopia Responses with red lens over right eye: ET patient sees two red lights to the **right** of three green lights. XT patient sees two red lights to the **left** of three green lights. RHT patient sees two red lights **below** three green lights. LHT patient sees two red lights **above** three green lights.

lights—two red and three green—represent diplopia. In the latter case, you must determine if it is crossed (XT) or uncrossed (ET), or vertically displaced (HT) diplopia.

Never ask the patient to describe what they see; it is too overwhelming a task. Ask specific questions that will get them talking and feeling as though they have control. For instance, the question "Do you see any lights?" requires a simple yes or no answer. Then ask, "Do you see any red ones/green ones?" and "How many of each?" If the patient reports five lights (three green, two red), you must then question very carefully to determine if diplopia is actually present, or if the patient is just rapidly alternating (alternate suppression). Specifically ask, "When you look at the red lights, do you still see the green lights off to the side, or *do they completely disappear?*"

When a child reports seeing all five lights all of the time, you must determine which kind of diplopia they are experiencing. Ask if the reds are on "this side," by tapping an arm or leg, or on the "other side," by tapping the other arm or leg. NEVER ask patients—no matter what age—to distinguish between right and left. They mix them up too often, and it is easier for you if they point to the side they mean. A child who reports three green and two red lights may actually have fusion, and is really seeing the bottom white light rapidly change from red to green. Though the patient reports five lights altogether, you know that one of the lights is just switching colors.

Objective fusion tests:
1. Cover-uncover test for presence of phoria or intermittent tropia
2. 4 diopter base-out test for suppression (or fusion)
3. Motor fusion amplitudes (convergence/divergence)

Once a patient's subjective fusion is established, you may rely on objective measurements of fusion. The *cover-uncover test* should then be done to determine the presence of a phoria, tropia or intermittent tropia. The *four diopter base-out test* may also be done at this time to test for suppression. Both the cover-uncover and four prism diopter base-out tests are objective tests of motor fusion. Because motor fusion maintains sensory fusion, its presence usually represents sensory fusion.

Orthoptists may want to check the subjective fusional status of a patient on a haploscopic device. It may be the only subjective fusion test that the child responds to and, because the eye position can be seen, the presence of harmonious and unharmonious ARC can be more easily determined.

Motor fusion amplitudes can be a combination of subjective responses and objective signs. Measure convergence last. You will be unable to do any testing that requires relaxation of the eye position or divergence *after* convergence testing (because convergence is so strong). For instance, if you need to measure the size of the eye turn, you must do it *before* any vergence testing, as vergence testing may alter the deviation that you measure.

Before measuring motor fusion under normal seeing conditions, determine if patients will be aware of diplopia when they reach their break point. Will they report diplopia accurately to you, or must you rely on watching their eye position closely while testing? To determine this, quickly check their ability to appreciate *physiological diplopia*. Instruct them to fixate on your nose and hold a pen or pencil up closely to their eyes. Ask how many pens they see. If they see two, they have no temporal retinal suppression under binocular conditions. Then hold the pencil about three feet from their eyes and ask them to fixate on it. Ask them how many noses they see. Again, they should see two, which would indicate no binocular nasal suppression. Because divergence has not yet been measured, be cautious that patients do not exercise convergence, which will change any succeeding measurements. (Do not allow them to converge any closer than three feet.)

A Risley prism or prism bar may be used to measure amplitudes. (However, if the Risley prism is used in the phoropter, it is impossible to monitor the patient's eye position.) Divergence is *always* measured before convergence, and distance measurements are *always* taken before the near measurements (see Table 6.1).

The patient is instructed to try to keep the 20/40 target letter single, but to report when it goes to double. Progressively larger base-in prism for divergence and progressively larger base-out prism for convergence are introduced before one eye. The **break point** is recorded as the prism where diplopia occurred and could not be overcome. Gradually smaller prism is then immediately introduced until the patient can fuse again. This is recorded as the **recovery point.** A patient who becomes diplopic at 25 P.D. and recovers at 18 P.D. would have the two points recorded as 25/18.

When convergence is measured, the break and recovery points are also recorded. In addition, the patient is requested to report when blurring occurs, in spite of single binocular vision. This blurring signals the use of accommodative convergence instead of fusional convergence. The prism power where this blurring occurs is recorded as the **blur point.** The maximum amount of base-out prism that can be immediately overcome is the **jump convergence,** and is measured by introducing a large base-out prism to one eye. Information gained from measuring jump convergence is questionable. However, in some patients a "poor" jump convergence is indicative of the patient's inability to efficiently change convergence from one fixation distance to another. (They have trouble looking up from a book to the black board, or vice versa.)

As a general rule, near measurements in P.D. are usually twice the amount of the distance measurements, and con-

Establish the patient's appreciation of physiological diplopia before testing motor fusion amplitudes.

A Risley rotary prism is two 15 P.D. prisms that can be rotated to negate each other equalling zero (placed base to apex), or rotated to equal any amount of prism power up to 30 P.D. (placed base to base).

Table 6.1 Order of Measurements

1. Distance divergence
2. Near divergence
3. Distance convergence
4. Near convergence

Table 6.2 Divergence/Convergence Norms

	Divergence	Convergence	Blur	Jump
Distance	6/4	14/12	no	12
Near	12/10	30/25	no	25

vergence measurements are twice the amount of divergence measurements. The recovery point is usually one notch down from the break point on the prism bar, and the jump point is ordinarily the same as the recovery point. Table 6.2 shows the norms for the various values. Keep in mind that these are norms and not "normal values." The patient's real convergence requirements determine if convergence is adequate, not where the patient falls in comparison to these norms. The convergence requirements for an accountant are usually quite different from those of a farmer. Whenever the possibility of convergence insufficiency is being considered, the patient's amplitudes must be judged—whether or not they are adequate for the patient's own needs.

Quantifying Strabismus

Quantifying Strabismus:
1. Distance, cc and sc, primary and secondary positions
2. Distance, with minus lenses if fusing accommodative ET
3. Distance, right and left tilt if vertical deviation
4. Near, with and without bifocal
5. Near, with plus lenses if ET
6. Near, with minus lenses if eso**phoric**
7. Prism and cover, simultaneous prism and cover if tropic
8. Hirschberg or Krimsky if necessary
9. Maddox rod measurements if cooperative

Once sensory fusion testing is completed (EXCEPT for motor amplitudes, the size of the eye turn may be measured without fear of breaking down the patient's deviation. Patients are to be measured while wearing their full correction. When a hyperopic correction is worn for correcting the eye turn, the deviation should be measured both with and without the correction. If a bifocal is worn to normalize a high AC/A ratio, measurements should be taken at near with and without the bifocal, to determine the need for it. Measurements should then be taken with minus lenses over the glasses to determine if a partial decrease in the plus correction can be tolerated. The decrease in plus correction can be safely prescribed if the patient remains phoric with the decrease.

Head tilt measurements are necessary on any patient with a vertical deviation. The prism must be held parallel to the floor of the patient's orbit.

Patients should routinely be measured with their correction on at distance, both in the primary position and in the secondary positions (up, down, right gaze, and left gaze). This may be done by turning the head approximately 25–30° (see Figure 6.4). If a vertical deviation is found or suspected, the tertiary positions (the corners) should be measured and measurements with the patient's head tilted to the right and to the left shoulder should be taken (see Figure 6.5). These tilt measurements provide the necessary information for the Bielschowsky three-step test (see Chapter 3) and should be done with the prism base held parallel to the floor of the patient's orbit—NOT parallel to the floor of the room.

Measuring in the secondary positions allows for detection

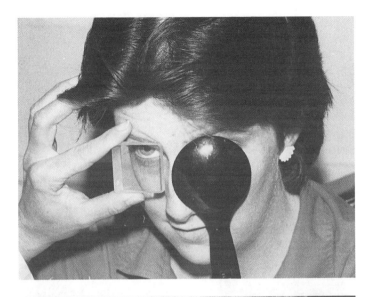

Figure 6.4 Prism and cover measurements in upgaze. Patient points chin down so eyes are in upgaze.

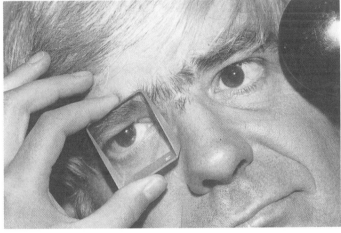

Figure 6.5 Prism and cover measurements during right head tilt. Prism is held so that the base of the prism is parallel to the floor of the orbit.

of an A or V pattern, or lateral gaze incomitance (CN VI palsy, Duane's syndrome, internuclear ophthalmoplegia). Because the young child must be approached as though each piece of information is the last that can be obtained, you must carefully decide which position of gaze you want to measure after the primary position. If the child has a history of a head turn to the right or left, measure right and left gaze BEFORE up and down gaze. The same applies if the child has a history of palpebral fissure narrowing when he looks to one side (Duane's syndrome). A child with a chin-up or chin-down head position should have up and down gaze measured BEFORE right and left gaze. Because V patterns are generally more common than A patterns, a child with an esotropia in primary should have DOWN gaze measured first (where the

Figure 6.6 Near fixation for infant or very young child using penlight and accommodative target for fixation.

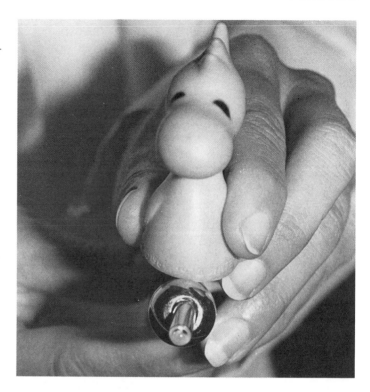

The light used during Hirschberg and Krimsky measurements is **not** primarily for the patient to fixate on. The "fixation" light is used so that the examiner can see!

Eye in:

Corneal reflex out

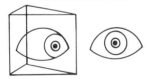

Correcting prism base out

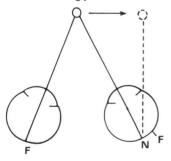

Image displaced out

Figure 6.7 RET results in displacement of corneal reflex out, correcting prism needs to be placed base out, and the image is displaced out (temporally).

deviation will be greatest with a V pattern). A child with an exotropia in primary should have UP gaze measured first. These are small items, but they can potentially save the uncooperative child from having to return for further examining.

Objective methods of measuring tropias on a very young child may be done by **Hirschberg** or by **Krimsky** measurements. Both tests involve near measurements, and although the corneal light reflex is assessed, *the child does NOT fixate on the light*. The light is only used so the *examiner* can see. As with any measurement test, accommodation should be controlled. Have the child fixate on an accommodative target. The light should be held close to the target so you can determine the child's eye position (see Figure 6.6).

Hirschberg Measurements

Hirschberg measurements assess the relative position of the light reflex in each eye. If the reflexes are symmetrical, even if nasally located (positive angle kappas, see Chapter 3), the child has no eye turn. When the light reflexes are assymmetrical, a tropia may exist. The direction of the eye turn is determined by the direction of displacement of the corneal light reflex from the normal position. An eye that deviates INward has a corneal reflex displaced OUTward. The correcting prism would be placed base-OUT and the diplopic image would be displaced OUT (see Figure 6.7). The amount of eye turn is the number of millimeter displacement where 1 mm displacement = 7° = 15 P.D. (approximately). Considering that a child's pupillary diameter is about 4 mm and the corneal diameter 12 mm, a light reflex at the pupillary margin indicates a 30 P.D. turn. A light reflex at the limbus indicates a 90 P.D. turn (see Figure 6.8)

Krimsky Measurements

Krimsky measurements use prisms to artificially place the deviated light reflex back to its original central position (or to the appropriate relative position when compared to the fixing eye). A child who fixates with either eye can have the prism placed over either eye. Correcting prisms are placed over the fixing eye in patients having a blind eye that cannot fixate, or with young children who strongly prefer one eye. In this way, you know that the eye under the prism is fixing correctly and are better able to see the light reflex of the deviating eye. A base-out prism is used to correct and ET, a base-in prism to correct an XT. A base-down prism is used to correct a hypertropia, and a base-up prism to correct a hypotropia.

Prism and Cover Measurements

Prism and cover measurements (P + C) require steady fixation with either eye and, therefore, cannot be used on infants or very young children. P + C measurements also require that both eyes move freely, so they cannot be used on patients with a restrictive strabismus such as Grave's ophthalmopathy. The patient must fixate on an accommodative target (a 20/40 or smaller letter) AND continue to accommodate on it. Non-accommodation leads to a major source of error when taking measurements. Instruct the patient to read different lines of letters—backward, forward, and to tell, for example, what the third letter of the second line is—so that the patient continues to accommodate. Do not instruct patients to simply "look at the E." Once they have seen it (regardless of its size), they can look at it steadily *without* necessarily accommodating. When this happens, the full deviation does not become manifest and this prevents accurate measurement.

The cover is alternately placed over each eye, NEVER allowing binocularity to take place. The direction of the phoria, tropia, or intermittent tropia has been determined by the cover-uncover test and the correcting prism is placed over either eye. The prism is changed until no movement of either eye occurs with cross-covering. The use of a larger prism results in reversal of eye movement upon cross-covering. Sometimes, a patient does not fixate well or steadily, and the eyes make an extra back and forth movement—a **redress** movement—each time the cover is moved. Neutraliza-

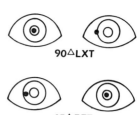

90△LXT

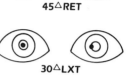

45△RET

30△LXT

Figure 6.8 Hirschberg Measurements: (top) 90 P.D. LXT: light reflex of the left eye deviates to the nasal limbus. (middle) 45 P.D. RET: light reflex of the right eye deviates to the middle of the temporal iris. (bottom) 30 P.D. LXT: light reflex of the left eye deviates to the nasal pupillary border.

Cover-uncover test determines the presence of a phoria or tropia. As the occluder is placed over one eye, the **opposite** eye is watched for movement. If movement occurs, a tropia of that uncovered eye is present. If no movement was noted, either a phoria, or no eye turn is present. The cover is then removed and that eye being uncovered is watched for movement. If that eye moves to pick up fixation either a phoria exists, or if a tropia had already been established during the initial "cover" part of the test, a strong eye preference for that uncovered eye exists. Then the opposite eye is covered and then uncovered. Each time, as the eye is covered, the **opposite** eye is watched for movement (indicating a tropia). As the eye is uncovered, **it** is watched for movement. If no tropia had been established and the uncovered eye moves to pick up fixation, a **phoria** is present.

If both eyes cannot be neutralized with the same prism (only one eye will neutralize with a particular prism), consider that there may be a primary and secondary deviation.

Upon covering, movement of the opposite eye **in**ward indicates that the eye had been **out** (exo). Movement of the eye **out**ward indicates that the eye had been **in** (eso). Movement of the eye **down**ward indicates that the eye had been **up** (hyper). Movement of the eye **up**ward indicates that the eye had been **down** (hypo).

Objective measurement tests:
1. Hirschberg
2. Krimsky
3. Prism and cover
4. Simultaneous prism and cover

tion is estimated when the movement in each direction is an equal amount.

Occasionally, only one eye will seem to be neutralized with one prism. This is a tip-off that a **primary and secondary deviation** exists (see Chapter 3). A patient who measures the *same* with either eye fixing will appear to be neutralized when the same prism is placed over either eye. When the deviation cannot be neutralized by only one prism, you must consciously measure the primary and secondary deviation.

First, cover the left eye so that the right eye is fixing. Place the correcting prism under the cover and over the deviated eye (the left eye). Then move the cover to the right eye and watch for movement of the *left eye*. With the appropriate correcting prism, no movement of that eye should occur when the cover is moved. Record the deviation with the right eye fixing.

Then measure the deviation with the left eye fixing. Cover the right eye and place the correcting prism under the cover. Move the cover from the right eye to the left and watch for movement of the right eye. This will be the deviation with the left eye fixing.

Simultaneous Prism and Cover

Another way of measuring a deviation with either eye fixing is by the **simultaneous prism and cover test** (S P + C). A patient is asked to fixate with one eye, and that eye is covered as the correcting prism is simultaneously placed over the deviating eye. (The cover and prism are simultaneously introduced.) If no movement is perceived, the deviation is neutralized. Because this method does not involve cross-covering, the S P + C only measures the tropia that is normally manifest. (Cross-covering measures the largest deviation possible and the size of the phoria also.)

Maddox Rod Measurements

Maddox rod measurements will measure a phoria or tropia, but require a cooperative, honest patient who has NRC and little or no suppression. Because it is a subjective test, the measurements are very accurate, and either eye can easily be made to fixate to measure a primary and secondary deviation.

The Maddox rod is a series of high-powered cylinders that are lined up together so that when held close to the eye, a white light will appear as an elongated line perpendicular to the direction of Maddox rod. Most Maddox rods are red, but white ones are available for use in trial frames for measuring torsion. Show the patient a *vertical line* when measuring a horizontal deviation and a *horizontal line* when measuring a vertical deviation. The rod AND correcting prism are

ALWAYS placed over the same eye (the deviating eye) and the patient is instructed to look at the light. The Maddox rod is a diplopia test, so a patient reporting uncrossed diplopia indicates and eso; a patient who reports crossed diplopia indicates an exo, etc. (see Chapter 2).

Patients are tested in the dark, and testing may be done at distance and at near, as well as in the nine positions of diagnostic gaze. The patient is told to "look at the white light and tell where the red line is." The Maddox rod is then placed over one eye for only a few seconds, and the patient is asked to decide where the red line is in relationship to the light. The eye *without* the Maddox rod is the fixing eye, and this should be recorded in the chart. The correcting prism is then placed with the Maddox rod over the non-fixing eye for a few moments, and the patient is asked if the red line is superimposed on the light. The prism size used when super-impositioning is accomplished is also the size of the deviation. Measurements may then be taken with the other eye fixing.

Double Maddox rod testing is a subjective test that quantifies the amount of torsion a patient experiences. Two Maddox rods, a red one and a white one, are placed in a trial frame with the orientation set to 90°. The patient then looks at a light in a completely dark room and is asked if both lines are parallel to the floor of the room. If they are, no torsion exists. If one line is tilted, you must determine if it appears **intorted** or **extorted** to the patient for that eye. The trial frame axis is then rotated until both lines are parallel to each other and to the floor. The number of degrees from 90° is recorded (see Chapter 3).

Maddox rod measurements:
1. Instruct patient to look at the **light.**
2. Introduce Maddox rod over one eye.
3. Record fixing eye (the one without the Maddox rod).
4. Assess where is the **line**? (ET, XT, HT, hypo)
5. Place correcting prism with the Maddox rod over the **non**fixing eye.

Hess Screen/Lancaster Red-Green Testing

The **Hess screen** and **Lancaster red-green test** are similar tests that map out the patient's subjective deviation while fixing with either eye. They are NOT diplopia tests, and they very easily show the primary and secondary deviation and over- and underactions of the EOMs (Hurtt, Rasicovici, Windsor, 1977).

Diplopia Fields

Diplopia fields are particularly useful in patients who have variable diplopia—patients with thyroid eye disease, for example. It is most easily done on the Goldmann perimeter with both eyes open and the patient's nose bridge centered. A III4e-size target is used, and the patient is instructed to follow the little light while holding his head still. When he sees double, he is to signal, and the area of double vision is mapped out. This is particularly useful when comparing pre-

Subjective measurement tests:
1. Maddox rod
2. Double Maddox rod (for torsion)
3. Hess Screen/Lancaster Red-Green Testing
4. Diplopia Fields

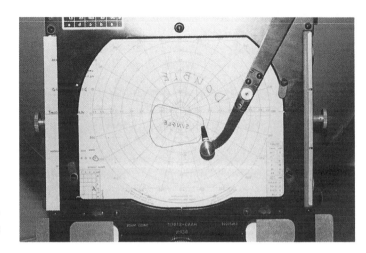

Figure 6.9 Diplopia field done on Goldmann perimeter showing area of single vision and double vision.

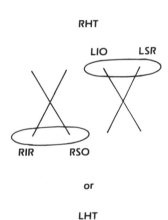

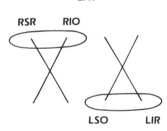

Figure 6.10 B3ST: Step 1. A RHT could be due to a palsy of the right eye depressors RSO, RIR or the left eye elevators LIO LSR. These two muscle pairs are circled to indicate them as possible causes of the incomitant RHT (top). A LHT could be due to a palsy of the right eye elevators RSR, RIO or the left eye depressors LIR, LSO (bottom).

and postoperative measurements or when trying to show patients where their area of useful single binocular vision is (see Figure 6.9).

Bielschowsky Three Step Test (B3ST)

The B3ST is based on the synergistic and antagonistic qualities of pairs of EOMs. There are three steps, each of which eliminates $\frac{1}{2}$ of the remaining potential muscles leaving only one muscle to be blamed after the three steps. The B3ST requires working knowledge of all muscle actions and in what position each action is best executed (see Chapter 2).

STEP 1: Determine the presence of a RHT or LHT in the primary position. If the patient's presenting sign is a *hypo*-deviation, consider it a *hyper* deviation of the opposite eye. A RHT implies that the weak muscles could be the right eye depressors (RSO, RIR) or the left eye elevators (LIO, LSR) (see Figure 6.10). A LHT implies that the weak muscles could be the right eye elevators (RIO, RSR) or the left eye depressors (LSO, LIR). The two possible muscle pairs are circled in Figure 6.10.

STEP 2: Determine if the HT is larger in right or left gaze. A HT largest in right gaze implies that the right gaze vertically acting muscles are weak. (The RSR, RIR work vertically in ABDuction—in right gaze; the LIO, LSO work vertically in ADDuction—in right gaze). The left gaze vertically acting muscles are the LSR, LIR, and the RIO and RSO (see Figure 6.11). Again, the two possible muscle pairs are circled. After Steps 1 and 2, when the circles are superimposed only two muscles have *two* circles around them.

After Step 2 only two muscles are left as possible candidates. There is ALWAYS: one from each eye; one is an oblique, the other is a rectus muscle, but *BOTH* are either superior muscles (intorters) or inferior muscles (extorters).

STEP 3: Determine if the HT is larger when measured during head *tilt* to the right or left. Proper measurement requires holding the base of the prism parallel to the floor of the *orbit*; not parallel to the floor of the room. The Maddox rod and correcting prism should be held so that the line and base are parallel to the floor of the orbit also (see Figure 6.12).

When the patient tilts to the right, the right eye attempts to "right" itself by intorting, the left eye tries to extort. The right head tilt torters are the RSO, RSR, LIO, LIR. The left head tilt torters are the RIO, RIR, LSO, LSR (see Figure 6.13). These two muscle pairs would be circled leaving only one muscle with three circles around it when the circles are superimposed. This is the palsied muscle.

After Step 2, two intorters *or* two extorters are left as possibly weak muscles. One is from each eye and only *one* is being tested during head tilt to the right while only the other is tested during head tilt to the left.

For example, if after Step 2, the RSR and LSO were left; both are superior muscles (intorters), one is a recti, the other an oblique, and one is from each eye. During head tilt to the *right*, only the intorters of the right eye are innervated. IF the RSR were the palsied muscle, its synergist for intorsion, the RSO would help intort. The RSO is also a depressor and working to depress the eye against a weak RSR would result in the eye dropping hypo. Thus, the original LHT would be largest during head tilt to the right in a RSR palsy.

If the LSO were palsied instead, it would try to intort the eye only during head tilt to the left. Its synergist for intorsion, the LSR, would also be *elevating* the eye against a weak LSO which should normally be depressing the eye.

Right gaze verticals

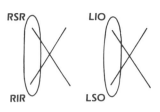

or

Left gaze verticals

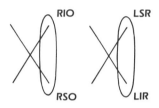

Figure 6.11 B3ST: Step 2. A HT worse in *right* gaze is due to a palsy of one of the "right gaze verticals": RSR, RIR, LIO, LSO. (Top) a HT worse in *left* gaze could be due to a palsy of one of the "left gaze verticals": LSR, LIR, RIO, RSO (bottom).

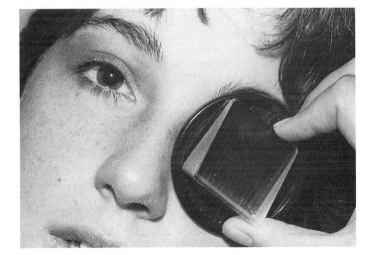

Figure 6.12 Maddox rod measurements of patient with RHT. Head is tilted to the right and base of prism is held parallel to floor of *orbit*. Maddox rod is held so that red line seen is parallel to floor of orbit also.

Right head tilt torters

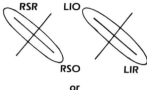

RSR   LIO

RSO   LIR

or

Left head tilt torters

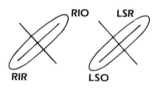

RIO   LSR

RIR   LSO

Figure 6.13 B3ST: Step 3. A HT worse on *right* head tilt could be due to a palsy of the right eye intorters: RSR or RSO, or the left eye extorters: LIO or LIR (top). A HT worse on left head tilt could be due to a palsy of one of the right eye extorters: RIR or RIO, or the left eye intorters: LSO or LSR (bottom).

A weak LSO would result in an increased LHT during head tilt to the left.

Circle the pairs of muscles potentially at fault after each of the three steps in the following examples.

Example:

> Step 1: LHT
> Step 2: Worse in left gaze
> Step 3: Worse in left tilt

Step 1 leaves: LSO, LIR, RIO, or RSR as possibly weak
Step 2 leaves: LIR, or RIO as still possibly weak
Step 3 leaves: RIO weak because the right eye *extorts* during left head tilt (see Figure 6.14).

Example:

> Step 1: RHT
> Step 2: Worse in right gaze
> Step 3: Worse in left tilt

Step 1 leaves: RSO, RIR, LIO, or LSR as possibly weak
Step 2 leaves: RIR, or LIO as still possibly weak
Step 3 leaves: RIR weak because the right eye *extorts* during left head tilt (see Figure 6.15).

Example:

> Step 1: small RHT, nearly ortho
> Step 2: RHT worse in left gaze, but LHT worse in right gaze
> Step 3: RHT worse in right tilt, but LHT worse in left tilt

This last example is a special B3ST problem. It is a classic example of a *bilateral SO palsy*, a fairly common finding following head trauma. Although there may be almost no deviation in the primary position, there is usually a large V pattern ET and RHT in left gaze and LHT in right gaze. The patient also may complain of torsional diplopia as bilateral SO palsies often result in over 10 degrees of torsion. To make the B3ST work in this situation, consider the RHT by itself (Step 1: RHT, Step 2: RHT worse in left gaze, Step 3: RHT worse in right head tilt) which will indicate a RSO palsy. Then consider the LHT by itself (Step 1: LHT, Step 2: LHT worse in right gaze, Step 3: LHT worse in left head tilt) which will indicate a LSO palsy (see Figure 6.16).

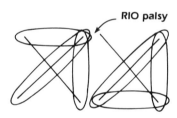

RIO palsy

Figure 6.14 Cyclovertical muscle pairs implied by B3ST in a RIO palsy.

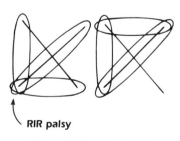

RIR palsy

Figure 6.15 Cyclovertical muscle pairs implied by B3ST in a RIR palsy.

Vision Testing and Refractive Error

Vision testing is covered in detail in Chapter 4 but a few comments are worthwhile here.

Monocular vision testing requires occlusion of one eye for a few minutes; this occlusion can dissociate the eyes. Once fusion testing is completed, vision testing may be done, but it should be done before any bright lights—such as for Hirschberg, Krimsky, version testing, slit lamp examination, or funduscopy—are exposed to the eyes that might create an after image.

Vision testing on a young child is a time-consuming and frustrating part of the exam. Unfortunately, because fusion must be assessed prior to occlusion, vision is often tested toward the end of the exam. Have the child wear his glasses and before patching the child with an on-face patch, decide which test can be successfully negotiated. Never assume that a 3-year-old child doesn't know his letters yet or that a 6-year-old child does. Be sure to record what test was done and say if a full line presentation was used.

Give parents copies of the Allen pictures for their child to take home and practice with before their next visit. Instruct the parents how to practice the E-game or the STYCAR matching test at home.

Measuring Vision with Nystagmus

Look for any nystagmus once you have patched the eye. If there is any nystagmus, use other alternate forms of occlusion. Standard opaque occlusion may bring out the latent nystagmus, so occlusion must be done in other ways. A high plus lens (+6 – +10) may be used to "occlude" the eye so that both eyes may be kept open with light coming in. Any vision recorded belongs to the eye without the lens as the vision with the lens is blurred.

Refractive Errors

A **fully balanced refractive error** is imperative for fusion development. Balanced means that any anisometropia is corrected so that regardless of how much the child must accommodate, both eyes send a clear image to the brain. An example of a balanced correction is as follows: Cycloplegic refraction OD +1.50 sphere, OS +5.00 sphere. Either the full correction could be given to each eye or if the full plus is not desired, each eye could be cut by an equal amount. If +1.50 is the amount to be cut, the right eye would be given plano, and the left eye would be given +3.50. In this way, when the fixing eye accommodates +1.50 in order to see clearly, and the fellow eye accommodates the same amount, both eyes would have the correct amount of plus correction to see clearly.

Anisometropia can be small and still cause a problem if unbalanced. Typically, after years of accommodative esotropia, a child who is fusing well with his correction on (i. e., OD +2.00 sphere, OS +1.25 sphere), can still hold his eyes straight without the correction during the exam. The temptation here is to remove the glasses altogether, but this leaves the right eye under-plussed by 0.75 diopters. The child will always fixate with the eye that requires less accommodation—the left eye—so the right eye will never receive a

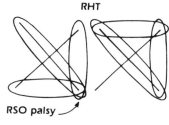
RHT

RSO palsy

and LHT

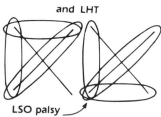
LSO palsy

Figure 6.16 Cyclovertical muscle pairs implied by B3ST in a bilateral SO palsy. (Top) RSO palsy; (bottom) LSO palsy.

A "balanced" correction is one that permits each eye to accommodate the exact **same** amount.
A "fully balanced" correction is one that does not permit any accommodation by either eye.

Side effects of cycloplegic
agents include:
1. Rash
2. Fever
3. Hallucinations
4. Rapid heart beat
5. Flushing of skin
6. Excitability

clear image. If the child has reached visual maturity, the harm is not permanent. Without correction, the child will never have perfect binocular vision because of the unbalanced refractive error.

Cycloplegics should be used in refracting young children. Cycloplegics are parasympatholytic drugs that paralyze accommodation and constriction. The side effects of cycloplegic agents used in young children are well known (1986 Physician's Desk Reference).

Children tend to over-accommodate when being refracted, and this changes the refractive error found. It will be less plus than truly exists. Because so many of these young patients will have presumed accommodative esotropia and therefore will have a hyperopic correction, it is imperative to find the total amount of hyperopia. The treatment of these children is to have them wear their *full* plus correction until they relax into the correction and fuse. It is absolutely necessary to use cycloplegics with these patients to give them the full plus.

While some clinicians feel confident using weak cycloplegics (mydriacyl or cyclogyl as opposed to atropine) or even no cycloplegic at all, any young child with an onset of esotropia that presumably is accommodative must be given the full plus correction. If the child returns after wearing the correction and is still esotropic, a full atropine refraction should be done to see if there is some additional plus that was missed. (So why not do an atropine refraction in the first place and save the parents an extra useless pair of glasses?) Children who are still atropinised when they go to pick up their new glasses will gladly wear them so that they can see. On the other hand, children whose cyclogyl wore off days before receiving the glasses will not be impressed by the hyperopic correction. It will not improve their vision but it will definitely hurt their ears and egos. It is important to be certain that the children are initially receiving their full cycloplegic refraction and that is best done with an atropine refraction.

Once an accommodative esotrope is fusing, the usual treatment is to gradually decrease the plus correction as long as the child can still maintain fusion. A phoria, no matter how large, indicates fusional control. To determine if a child's correction can be decreased, use minus lenses over his own correction and measure the fusion and deviation with the cut prescription. This keeps the child balanced. Never cyclolplege one of these children, because they do not need to be. Once children are fusing with a particular balanced correction, discovering how much *more* hyperopic they are does not matter—it only matters if they are fusing and *balanced.* So cycloplegling them will not add any information and will only stimulate them to try to accommodate while they are cyclopleged. That may cause the eye to cross in for a period of time unnecessarily.

Examples of the Systematic Approach

A tentative diagnosis is formed from the history. Certain tests that will confirm or disprove the diagnosis should be done, and in the proper order because each test will influence the tests that follow.

Infantile ET

Document equal vision before shining any lights in the baby's eye. Document ABDuction of each eye. Look for overactive IOs and the presence of a DVD. Measure the size of deviation and then refract.

Exodeviations

Measure stereopsis, sensory fusion (W4D), fusional control (cover-uncover); document suppression and retinal correspondence if XT. Measure the size of the deviation including downgaze reading position, and document the amplitudes (divergence and convergence). Record visual acuity including near vision and NPA.

Accommodative ET

Measure stereo and fusion with glasses on, fusional control with correction and with minus lenses if fusing or with additional plus if ET. Amblyopia is frequent so assess visual acuity carefully. Determine if near deviation is greater and if it is, if it is due to a high AC/A ratio, measure with plus adds. If a V pattern is suspected, look for overactive IO and for incomitance at *distance*. Do a cycloplegic refraction on the initial visit or if the distance deviation is not adequately controlled at subsequent visits.

Congenital XT

Health and visual acuity of the deviated eye is most important. Documentation of ADDuction of the deviated eye must be shown. Measure the size of the deviation and look for incomitance.

CN VI Palsy

Document ABDuction (or lack of) if possible, fusion with head position yielding binocularity, visual acuity of deviated eye if young enough to develop amblyopia. Document incomitance and primary/secondary deviations with P + C.

Abnormal Head Positions

Document stereo and/or fusion with head position. Measure eye position with head straightened, occlude one eye to see if the head straightens automatically. Look for ptosis and A or V patterns if there is a chin up or chin down head position. Measure deviation to determine presence of muscle palsy. Use Double Maddox rods if there is a head tilt.

References

Hurtt, J., Rasicovici, A., Windsor, C. E.: Comprehensive Review of Orthoptics and Ocular Motility. 2nd ed. Saint Louis, Missouri: C.V. Mosby 1977, p 98.

1986 Physician's Desk Reference. 40th ed. Oradell, New Jersey: Medical Economics Company Inc., 1986.

Tillson, G.: Two new clinical tests for stereopsis. *American Orthoptic Journal*, 35: 126–134, 1985.

CHAPTER 7

Differential Diagnosis

Introduction

The **differential diagnosis** refers to the thought process that leads an examiner through the exam to a final diagnosis. The history provides the clues to set up the tentative diagnosis and the examination tests either to prove or disprove the tentative diagnosis. Gradually, through the process of elimination, or simply on the basis of test results, the final diagnosis is made by the physician. OMAs must know the technique of using a differential diagnosis so that the appropriate tests are done in the proper sequence.

The "differential diagnosis" varies depending on the age of the patient and on the tentative diagnosis. For instance, a sudden crossed eye in a two-year-old child is probably caused by something different from a sudden crossed eye in a 64-year-old diabetic. Various additional history questions and exam tests will help reach a final diagnosis.

The following tables list the main possibilities in order of likelihood, additional history questions, pertinent tests, and what the test results indicate.

Categories of Differential Diagnosis

Decreased *AB*Duction	Table 7.7
Decreased *AD*Duction	Table 7.8
Congenital findings	Table 7.11
Diplopia	Table 7.10
Decreased Ductions	Table 7.5
Decreased Elevation	Table 7.6
Enophthalmos	Table 7.16
Esotropia	Table 7.1
Exotropia	Table 7.2
Exophthalmos	Table 7.15
Abnormal *Head* positions	Table 7.9
Hypertropia	Table 7.3
Hypotropia	Table 7.4
Near vision problems	Table 7.12
Pseudostrabismus	Table 7.17
Decreased binocular *Vision*	Table 7.14
Decreased monocular *Vision*	Table 7.13

Table 7.1 Esodeviations

Tentative Diagnosis:	
1. Infantile ET	
a. Additional history:	
Onset before age six months	
Fixates with either eye	
Constant ET	
b. Test:	Indicates:
ABDuction	Reluctant but intact
Measurements	Large ET, V pattern
Cross-fixates	Equal vision
Cycloplegic refraction	Small refractive error
Versions	Possible IO overaction
Cover-uncover	Possible DVD
2. Accommodative ET	
a. Additional history:	
E(T), N>D	
Eye preference	
Worse than when tired, concentrating	
Typical age of onset 18 months–3 years old	
Range of onset 7 months–10 years	
b. Test:	Indicates:
P+C, cc+sc	Accommodative component
Cycloplegic refraction	<4 D hyperopia, high AC/A
	>4 D hyperopia, normal AC/A
Fusion, cc	Excellent usually
Vision	Amblyopia frequently
3. LR palsy (CN VI)	
a. Additional history:	
Severe head/neck trauma	
ET greater to right gaze, left gaze, or both sides	
Head position to right or left	
Horizontal, diplopia, variable, worse at distance	
Sudden onset of ET/diplopia	
Present since birth if congenital	
Associated with facial nerve palsy (Moebius or brain stem tumor)	
Associated with ear pain (Gradenigo)	
Worsens with fatigue (myasthenia gravis)	
b. Test:	Indicates:
Fusion with head position	Recent onset
Measurements	ET>right and/or left gaze
	ET D>N, slight A tendency
Measure with OD, OS fix	Primary/Secondary deviation
Versions/ductions, saccades	Decreased LR function
Diplopia testing	Horizontal, incomitant
	Young child may suppress
Vision	Possible amblyopia
Forced ductions	Negative
4. Duane's retraction syndrome type 1	
a. Additional history:	
No trauma	
Lid fissure changes with gaze	
Abnormal head position	
Present since birth	
b. Test:	Indicates:
Versions, ductions	No ABDuction, lid fissure narrows during ADDuction
Fusion with head position	Fusion often
Vision	May have amblyopia

Table 7.1 Esodeviations—Continued

5. Consecutive ET
  a. Additional history:
    Previous surgery for XT
  b. Test: Indicates:
    Fusion with ET corrected   Fusion usually if X or X(T) pre-op
6. Nystagmus compensation syndrome
  a. Additional history:
    Any nystagmus seen during ABDuction
    Large ET present since birth
    Cross-fixates turning head to see
  b. Test: Indicates:
    Krimsky   Over-converges with "correcting
                                          prism"
    Versions/ductions   Jerk nystagmus in ABDuction, o.u.
    Cross-fixation   Does *not* move fixing eye out of
                                          ADDucted position to mid-line
7. Cyclic ET
  a. Additional history:
    24-, 48-, or 96-hour schedule of alternation between ET one cycle and
      straight next cycle
    Fairly sudden onset during childhood
  b. Test: Indicates:
    ET day: Fusion   May have diplopia initially, then
                                          suppression/ARC
            P + C   Large ET, comitant usually
    Straight day: Fusion   Excellent
                P + C   Ortho or small eso*phoria*
                Vision   May have amblyopia
8. Strabismus fixus
  a. Additional history:
    Longstanding ET
  b. Test: Indicates:
    Forced ductions   Positive MR tightening
    Measurements   Large ET, greater in R and L gaze
    Versions/ductions   Decreased ABDuction, fixed
                                          ADDuction
    Vision   Must cross-fixate and turn head to
                                          see
9. Divergence paralysis
  a. Additional history:
    General health—NO raised intracranial pressure
    Infections, parasites, and travel abroad
    Trauma possible
  b. Test: Indicates:
    Divergence amplitudes   Nearly nonexistent
    Versions/Ductions   FULL, ABDuction okay
    Diplopia testing   Uncrossed, worse at distance
    Measurements   Comitant esodeviation D > N
    Fusion   May fuse at near
    Vision   Equal
10. Accommodative effort syndrome
  a. Additional history:
    Near asthenopia, blurring, or diplopia
  b. Test: Indicates:
    NPA   Normal
    P + C   E′, possibly E(T)′
    Divergence amplitudes   Poor
    Plus lenses for near   Help relieve symptoms

Table 7.1 Esodeviations—Continued

11. Pseudo ET
    a. Additional history:
       Onset — usually since birth
       Incomitance
    b. Test: Indicates:
       Hirschberg/Cover-uncover   No deviation
       External   Epicanthal folds frequently
       Angle kappa   Negative
       Fusion   Excellent, no suppression
       Versions/ductions   Appears to have increasing ET to
                                            right and left gaze

Table 7.2 Exodeviations

1. Basic, divergence excess type, convergence insufficiency type
   a. Additional history:
      Intermittency
      Worse at distance or near
      Worse with fatigue, illness
   b. Test: Indicates:
      Fusion   Usually excellent
      Measurements   Exodeviation, usually comitant
      Vision   Equal
      Convergence amplitudes   Poor fusional convergence amps
2. Infantile XT
   a. Additional history:
      Present since birth
      Eye preference
   b. Test: Indicates:
      Vision   Decreased in nonpreferred eye
      Health of eyes   No specific disease of eye
      Measurements   Usually large, comitant XT
3. MR palsy/CN III palsy
   a. Additional history:
      Present since birth if congenital
      Severe trauma
      General health: possible headaches, myasthenia gravis
      Other signs: ipsilateral ptosis, mydriasis, cycloplegia, hypo-
         deviation, contralateral body paralysis (Benedikt's syndrome)
   b. Test: Indicates:
      Versions/ductions   Decreased ADDuction
      Measurements   Incomitant exodeviation, greater
                                            in ADDuction
      Measure with OD, OS fix   Primary and secondary deviation
      Vision   Possible amblyopia, cycloplegia
      Head position   Present if to attain fusion
4. Duane's retraction syndrome type 2
   a. Additional history:
      No trauma
      Lid fissure changes with gaze
      Abnormal head position
      Present since birth
   b. Test: Indicates:
      Versions/ductions   Decreased ADDuction with lid
                                            narrowing, ABDuction okay
      Head position   Done to achieve fusion
      Fusion with head position   Often present
      Vision   May be amblyopic

Table 7.2 Exodeviations—Continued

5. Blind eye
   a. Additional history:
   Age of blindness in one eye (if over 7, likely to go exo)
   Constant non-alternating XT

   b. Test: Indicates:
   Vision   Blind eye
   Krimsky measurements   Fairly comitant XT

6. Consecutive XT
   a. Additional history:
   Previous surgery for ET

   b. Test: Indicates:
   Measurements   XT
   Fusion   Diplopia or Suppression/ARC

7. Cranial-facial anomalies
   a. Additional history:
   Apert syndrome (Harley, 1975)
   Crouzon syndrome (Harley, 1975)

   b. Test: Indicates:
   Measurements   Large XT, bilateral SO palsies
                                     often
   Versions/ductions   IO overaction, SO underaction
   External   Bilateral exophthalmos
                                     (Crouzon)
   Refraction   Astigmatism (Apert)
                                   Progressive hyperopia (Crouzon)

8. Convergence Paralysis
   a. Additional history:
   Trauma
   General health: recent neurological condition possible

   b. Test: Indicates:
   Convergence amplitudes   Nearly nonexistent
   Measurements   Comitant exodeviation N > D
   Versions/ductions   Full
   Diplopia   Crossed, worse at near
   Fusion   Distance only
   Vision   Equal

9. Internuclear Ophthalmoplegia (INO)
   a. Additional history:
   Bilateral: Systemic multiple sclerosis
   Unilateral: Vascular accident, inflammation, infection, or
     tumor in brainstem, possibly myasthenia gravis

   b. Test: Indicates:
   Versions/ductions   Decreased ADDuction with jerk
                                     nystagmus in ABDucted eye
   Convergence amplitudes   Intact convergence

10. Pseudo XT
    a. Additional history:
    Constant

    b. Test: Indicates:
    Krimsky/Cover-uncover   No eye deviation
    Fusion   Excellent, no suppression
    Angle kappa   Positive
    Funduscopy   May have retinopathy of pre-
                                     maturity with temporally
                                     dragged fovea

Table 7.3 Hyperdeviations

1. Isolated CN palsy (SO most common)
   a. Additional history:
      Trauma, may be mild
      Head tilt
      Combined horizontal and vertical diplopia
      Diplopia/asthenopia worse to right and left
      Myasthenia gravis

b. Test:	Indicates:
Fusion with head position	Usually fuses
P + C, 9 positions	Incomitant HT
B3ST	Isolates EOM palsy
Versions/ductions	SO: May show underactive SO, overactive IO, inhibitional palsy of contralateral (IPC) SR
	IO: May show underactive IO, overactive SO, IPC IR
	SR: May show underactive SR, overactive IR, IPC SO
	IR: May show underactive IR, overactive SR, IPC IO
Vertical amplitudes	May exist if congenital/longstanding
	No vertical amps if recent onset
Subjective torsion	May exist if recent onset
	No torsion if congenital/longstanding
	SO, SR: extorsion
	IO, IR: intorsion
Look at old photographs	Old head tilt if congenital/longstanding

2. DVD
   a. Additional history:
      Associated with congenital/infantile ET
      One or both eyes seen to go up, neither eye ever goes hypo
      Intermittent HT, either eye

b. Test:	Indicates:
Cover-uncover	Either eye elevates under cover without associated hypodeviation of the fellow eye
P + C	Difficult to measure, variable
Versions/ductions	Rule out IO overaction as "cause"

3. Brown's syndrome
   a. Additional history:
      Trauma to trochlear region of globe
      Sinus, orbital surgery
      Juvenile rheumatoid arthritis
      Present since birth if congenital

b. Test:	Indicates:
Versions/ductions	No elevation of globe in ADDuction
	Eye elevates easily in ABDuction
Forced ductions	Restriction of globe up and in
Fusion	Often fuses in downgaze
Vision	Amblyopia may or may not be present
Krimsky	Hypotropia of affected eye when up and in

Table 7.3 Hyperdeviations—Continued

4. Blowout fracture
   a. Additional history:
   Blunt trauma—orbital fracture
   Diplopia/discomfort worse in upgaze often
   b. Test: Indicates:

Ductions/versions	Hypotropia worsens towards upgaze
	Restrictions of ductions: upgaze, maybe down, right, or left gaze
Forced ductions	Restriction of globe, upgaze usually
X-ray/CT scan	Orbital fracture, floor frequently, or nasal wall
Diplopia field	Often has region of single binocular vision
Exophthalmometry	Often affected eye enophthalmic

5. Double elevator palsy (DEP)
   a. Additional history:
   Present since birth
   Ptosis on affected side
   Chin up head position
   NO trauma
   b. Test: Indicates:

Fusion with chin up	Possible fusion
Versions/ductions	Constant hypotropia of affected eye
Forced ductions	No restrictions of elevation unless very longstanding
External	Pseudo ptosis on affected side

6. Grave's ophthalmopathy
   a. Additional history:
   Thyroid dysfunction in past or present
   b. Test: Indicates:

Ductions/versions	Restriction of upgaze (IR), lateral gaze (MR), or any EOM
Forced ductions	Positive for restriction
Fusion with head position	Usually has fusion
Diplopia	Variable, vertical > horizontal

7. Pseudo HT
   a. Additional history:
   b. Test: Indicates:

Cover-uncover	No deviation
Lift lid	Appearance of HT disappears
Assess pupils	Possible asymmetry
Document with photos	No deviation by light reflex

Table 7.4 True Hypodeviations (see Chapter 3)

1. Brown's syndrome—see Table 7.3
2. Double elevator palsy—see Table 7.3
3. Blowout floor fracture—see Table 7.3
4. Restrictive thyroid eye disease (Grave's)—see Table 7.3

Table 7.5 Decreased Ductions

1. Duane's retraction syndrome type 1 — see Table 7.1
2. Duane's retraction syndrome type 2 — see Table 7.2
3. Duane's retraction syndrome type 3
 a. Additional history:
  No trauma
  Lid fissure changes with gaze
  Present since birth

b. Test:	Indicates:
Versions/ductions	Absent ABDuction AND ADDuction with narrowing of lid fissure, enophthalmos during attempted ADDuction

4. Cyclovertical muscle palsy — see Table 7.3
5. Brown's syndrome — see Table 7.3
6. Blowout fracture — see Table 7.3
7. Thyroid ophthalmopathy — see Table 7.3
8. Strabismus fixus — see Table 7.1
9. Double elevator palsy — see Table 7.3

Table 7.6 Decreased Elevation

1. Thyroid ophthalmopathy — see Table 7.3
2. Brown's syndrome — see Table 7.3
3. Blowout floor fracture — see Table 7.3
4. Double elevator palsy — see Table 7.3
5. Myasthenia gravis — see Table 7.1
6. CN III palsy
 a. Additional history:
  Ptosis, mydriasis, decreased elevation AND ADDuction
  Trauma
  Present since birth if congenital
  Diabetes in adult

b. Test:	Indicates:
Versions/ductions	Absent elevation, ADDuction
Forced ductions	No restrictions unless very longstanding
Lid assessment	Ptosis, may be complete
Pupil assessment	No reaction to light, accommodation, unless diabetic
Vision/refraction	Cyclopleged, unless diabetic
Krimsky	Incomitance, often never ortho
ABDuction (LR)	Intact
Depression, intorsion (SO)	Intact

Table 7.7 Decreased ABDuction

1. Duane's retraction syndrome type 1 — see Table 7.1
2. CN VI palsy — see Table 7.1
3. Infantile ET — see Table 7.1
4. Nystagmus compensation syndrome — see Table 7.1
5. Strabismus fixus — see Table 7.1
6. Moebius syndrome — see Table 7.1 (LR palsy)
7. Duane's retraction syndrome type 3 — see Table 7.5
8. Myasthenia gravis — see Table 7.1

Table 7.8 Decreased ADDuction

1. Duane's retraction syndrome type 2 — see Table 7.2
2. CN III palsy, MR palsy — see Table 7.2, 7.6
3. Internuclear Ophthalmoplegia (INO) — see Table 7.2
4. Myasthenia gravis — see Table 7.1

Table 7.9 Abnormal Head Position

1. Congenital torticollis
   a. Additional history:
      Present since able to hold head up unassisted (approximately
         6 months)
      No eye turn ever seen, no nystagmus ever seen

b. Test:	Indicates:
Prolonged occlusion	Head position PERSISTS
EOM exam	No motility disturbance
Force head tilt to opposite side	NO hyperdeviation seen (No SO palsy)

2. Congenital nystagmus
   a. Additional history:
      Horizontal jerk nystagmus seen WITHOUT head position
      Nystagmus dampens during near fixation
      Present since birth

b. Test:	Indicates:
Binocular, monocular vision	Usually better binocular
Versions/ductions	Nystagmus worse away from null point
Assess nystagmus, D + N	Usually less at near

3. CN IV palsy — see Table 7.3
4. Duane's retraction syndromes types 1, 2, 3 — see Tables 7.1, 7.2, 7.5
5. Nystagmus compensation syndrome — see Table 7.1
6. Strabismus fixus — see Table 7.1
7. Grave's ophthalmopathy — see Table 7.3

Table 7.10 Diplopia

1. Uncorrected refractive error causing blur
   a. Additional history:
      Double vision very close together
      Possibly more than two images per eye
   b. Test: Indicates:
      Refraction   Astigmatism frequently, blur
                                               disappears
2. Monocular diplopia
   a. Additional history:
      Diplopia PERSISTS in one or both eyes with occlusion
      Extra image(s) may be blurred or "smeared"
      Injury to cornea/lens causing scarring
      Retinal "injury," laser
      Bifocal line obstructing visual axis
   b. Test: Indicates:
      Monocular occlusion   Diplopia persists
      Health of eye   Usually indicates organic cause of
                                               diplopia
                                               Rule out cataract or corneal
                                               opacity

True Binocular Diplopias
3. Decompensated phoria now tropic
   a. Additional history:
      Intermittent tropia with diplopia previously
   b. Test: Indicates:
      Fusion with deviation corrected   Fusion usually
      P + C   Often comitant, usually XT or HT
      Fusion potential   Usually excellent
4. EOM palsy—see Tables 7.1, 7.2, 7.3
5. Newly noticed diplopia from longstanding condition
   a. Additional history:
      Brown's syndrome—see Table 7.3
      Duane's retraction syndromes types 1, 2, 3—see Tables 7.1, 7.2, 7.5
      Thyroid ophthalmopathy—see Table 7.3
      Blowout fracture—see Table 7.3
      Have they ever looked into that field of gaze before? (Probably not.)
6. Secondary to restrictive strabismus
   a. Additional history:
      Thyroid ophthalmopathy—see Table 7.3
      Brown's syndrome—see Table 7.3
      Duane's retraction syndromes types 1, 2, 3—see Tables 7.1, 7.2, 7.5
      If the strabismus had been present since childhood, have they ever
         looked into that field of gaze? (Probably not)
7. Divergence paralysis—see Table 7.1
8. Convergence paralysis—see Table 7.2
9. Intractable diplopia
   a. Additional history:
      Previous anti-suppression exercises
      Trauma
      Strabismus surgery
   b. Test: Indicates:
      Fusion potential   None
      Occlusion   No diplopia
      Retinal correspondence   Often abnormal, paradoxical

Table 7.10 Diplopia—Continued

10. Paradoxical diplopia	
a. Additional history:	
Previous strabismus surgery	
Consecutive tropia	
b. Test:	Indicates:
Retinal correspondence	ARC
11. Glasses induced	
a. Additional history:	
New glasses	
New bifocal	
Aphakic correction	
Glasses recently adjusted	
Diplopia disappears sc	
Diplopia is typically vertical but may be uncrossed	
b. Test:	Indicates:
Check for prism in glasses	Optical center misalignment creating prism, typically vertical or base IN
Check height of bifocal	May be asymmetric

Table 7.11 Congenital Findings

1. Brown's syndrome – see Table 7.3
2. Duane's retraction syndromes – types 1, 2, 3 – see Tables 7.1, 7.2, 7.5
3. Double elevator palsy – see Table 7.3
4. Congenital syndrome – see Table 7.9
5. Moebius syndrome – see Table 7.1 (LR palsy)
6. Congenital EOM palsy – see Tables 7.1, 7.2, 7.3
7. Systemic syndromes: Apert, Crouzon, Down (Harley, 1975)
8. Congenital myasthenia

Table 7.12 Near Vision Problems

1. Convergence insufficiency	
a. Additional history:	
Blurring while reading/near work or prolonged distance fixation	
Occasional horizontal diplopia	
Difficulty changing from N to D fixation or D to N	
Headaches after using eyes; NEVER upon awakening	
b. Test:	Indicates:
Fusion	Normal stereopsis
P + C	May have small to large exo, eso, or be ORTHO
	Often has congenital SO palsy
Fusional amplitudes	Poor fusional amplitudes for visual demands, uses accommodative convergence, poor recovery/jump point
Prolonged occlusion	Symptoms completely disappear
2. Presbyopia	
a. Additional history:	
40+ years old	
Previous hyperopic correction	
Older myope who recently was first time fit with contact lenses	
Older hyperope who was recently switched from contact lenses to full time glasses	

Table 7.12 Near Vision Problems—Continued

b. Test:	Indicates:
Cycloplegic refraction	Possibly over-minused or latent hyperopia
NPA, cc	Should be normal for age

3. Systemic convergence insufficiency
  a. Additional history:
  Trauma
  Illness: encephalitis, drug intoxication, mononucleosis (Raskind, 1976)
  Increasing diplopia and blurring with near vision

b. Test:	Indicates:
As for convergence insufficiency	Convergence insufficiency
NPA	Severely decreased and often fixed
Plus lenses	Improves near vision
Near P + C	Constant XT at N requiring BI prism for fusion

4. Convergence paralysis—see Table 7.2
5. Divergence paralysis—see Table 7.1
6. Accommodative spasm
  a. Additional history:
  Possible psychogenesis
  Severe distance blurring after near fixation

b. Test:	Indicates:
Distance VA	Worse than 20/200 usually
Manifest refraction	Up to 8-10 D myopia
Cycloplegic refraction	High myopia disappears

7. Accommodative effort syndrome—see Table 7.1
8. Juvenile presbyopia
  a. Additional history:
  Drug use
  History of hysteria
  Symptoms of presbyopia except age is much younger

b. Test:	Indicates:
Near vision	Decreased
P + C	Insignificant eye turn
NPA	Decreased for age
Plus lenses for near	Relieves symptoms

9. CN IV palsy (SO palsy)—see Table 7.3
  a. Additional history:
  Asthenopia, diplopia increases in reading position

Table 7.13 Nonorganic Monocular Decreased Vision

1. Amblyopia
  a. Additional history:
  Strabismus
  Anisometropia

b. Test:	Indicates:
Vision	Unexplained difference in acuity
Vision with single optotypes	May be better than full line acuity
Fusion	Worse than bifoveal fusion
Four Diopter Base Out Test	Suppression

Table 7.13 Nonorganic Monocular Decreased Vision—Continued

2. Malingering
   a. Additional history:
      How is school, how is life?
      Pending litigation

   b. Test: Indicates:
      VF at 2 distances   "Cylinder" of vision, not "cone"
      VEP   Normal amplitudes and latencies
      Four Diopter Base Out Test   NO suppression
      Visual acuity   Inconsistent answers

Table 7.14 Nonorganic Decreased Binocular Vision

1. Bilateral "amblyopia"
   a. Additional history:
      Previously uncorrected high refractive error

   b. Test: Indicates:
      Vision   Decreased both eyes
      VEP   Decreased amplitudes and
                                           prolonged latencies usually
      Refractive error   High cylinder, or sphere
2. Malingering—see Table 7.13

Table 7.15 Exophthalmos

1. Grave's ophthalmopathy
   a. Additional history:
      Proptosis NOT present years ago
      Thyroid disease
      Diplopia, particularly in up gaze
      May be BOTH eyes but asymmetrical

   b. Test: Indicates:
      Exophthalmometry   Either greater than 22 mm or dif-
                                           ference between eyes > 2 mm
      Lid fissure height   May be greater on side of proptosis
2. Orbital tumors: Lymphoma, malignant melanoma, rhabdomyosarcoma,
      retinoblastoma, neurofibroma, glioma, dermoid, lacrimal gland
      tumor, carcinoma, mucocele
   a. Additional history:
      May be sudden onset
      May be any age, frequently children
      Associated with other illness

   b. Test: Indicates:
      Exophthalmometry   Exophthalmos
      CT scan, B scan   Orbital mass
3. Inflammations: Pseudo tumor, myositis
   a. Additional history:
      Pain
      Ophthalmoplegia

Table 7.15 Exophthalmos—Continued

4. Infections: Orbital cellulitis
   a. Additional history:
      Sudden onset
      "Hot" eye
      Monocular
      Young child
   b. Test: Indicates:
      CT scan   Orbital involvement
      General health   Sick child
5. Vascular disorders: Orbital varix, cavernous sinus thrombosis, pulsating
      exophthalmos
   a. Additional history:
      Eye bulges when baby cries (varix)
      Previous orbital infection (thrombosis)
      Neurofibromatosis
   b. Test: Indicates:
      Orbital venography   A-V malformation
      View patient from side   See pulsating exophthalmos
6. Orbital anomalies: Crouzon syndrome, Apert syndrome (Harley, 1975)
7. Enophthalmos of contralateral eye
   a. Additional history:
      Blowout fracture
      Blind, phthisical eye
   b. Test: Indicates:
      Exophthalmometry   Enophthalmos of contralateral eye
8. Pseudo Exophthalmos
   a. Additional history:
      Old photographs
      Unilateral lid retraction or ptosis
      Large eye due to buphthalmos (juvenile glaucoma), axial myopia,
         staphyloma
      Real enophthalmos of contralateral eye
   b. Test: Indicates:
      Exophthalmometry   Normal

Table 7.16 Enophthalmos

1. Blowout fracture—see Table 7.3
2. Phthisis bulbi—see Table 7.15
3. Pseudo—Real exophthalmos of contralateral eye—see Table 7.15

Table 7.17 Pseudostrabismus

1. Prominent epicanthal folds (ET) — see Table 7.1	
2. Narrow lateral canthi (XT)	
3. Angle Kappa: Positive (XT) — see Table 7.2	
Negative (ET) — see Table 7.1	

4. Ectopic macula
   a. Additional history:
      Previous retinal problems or surgery

b. Test:	Indicates:
Funduscopy	Displaced macula

5. Anisocoria
   a. Additional history:
      May be associated with mild ptosis with miotic pupil (Horner's syndrome)
      May or may not have heterochromia (congenital Horner's syndrome has heterochromia)
      May or may not have facial anhydrosis (drying of face)
      Trauma
      CN III palsy

b. Test:	Indicates:
Pupil evaluation	Anisocoria
Cover-uncover	No eye deviation
Fusion	Normal

6. Exophthalmos — see Table 7.15

References

Harley, R. D.: *Pediatric Ophthalmology*. Philadelphia, Pennsylvania: W.B. Saunders Company, 1975.

Raskind, R. H.: Problems at the reading distance. *American Orthoptic Journal*, 26:53–59, 1976.

CHAPTER 8

Treatment

The OMA's Role in Treatment

Patient compliance is much higher when patients thoroughly understand their eye problem, the treatment involved, and the prognosis. Patient understanding also increases the likelihood of treatment compliance. Misunderstanding strabismus has no socio-economic or racial barriers; it is confusing to everyone. Patients frequently are uncomfortable repeatedly asking their doctor to explain their condition again and the physician frequently complicates matters by being busy and resorting to technical language.

The explanation of the patient's problems is often best carried out by the ophthalmic medical assistant who has both the understanding and the time to spend with the patient. The patient invariably asks, "What did he mean?" right after the doctor has thoroughly explained the situation to the patient and left the room. As an OMA, you can be available to the patient for answers and be someone who is only a phone call away.

Goals of Treatment

Orthoptists today function as ocular motility specialists both evaluating and diagnosing problems and recommending and carrying out treatment. The treatment recommended may be surgical, spectacle, prismatic, pharmacological or truly orthoptic.

Ultimately, every patient with fusion potential should have *comfortable single binocular vision* (fusion). This requires equal, clear vision, and adequate motor fusion and sensory fusion for their visual needs.

When sensory fusion is not possible, motor fusion is not necessary. In these patients, *equal vision* is the primary goal of treatment. Even though only one eye is used at a time, it is important to have two good seeing eyes so that if one eye is damaged by trauma or disease, the other eye is a good "spare." People with severe amblyopia in one eye are not immune to devastating ocular accidents or subsequent disease in their good eye.

When equal vision is no longer possible, *good cosmesis* is the only goal left for the strabismic patient.

There are five main areas of treatment for the patient with an ocular motility problem.

Orthoptics

The word orthoptics means "straight eyes." Orthoptic treatment generally is aimed at amblyopia therapy, sensory fusion, or motor fusion therapy.

Occlusion therapy is used for treatment of amblyopia, diplopia, during recovery of an EOM palsy, and for anti-suppression or treatment of ARC.

Treatment of amblyopia by direct and partial occlusion is covered in detail in Chapter 4. Other aspects of occlusion therapy for amblyopia to be considered involve press-on prisms or lenses. Because these lenses blur vision somewhat, they can be used over the preferred eye to help "occlude" it when the lenses are being used to treat another aspect of strabismus. For instance, a young treatable mild amblyope who needs correcting prism for consecutive ET, could have the major part (or all) of the prism placed over the preferred eye thus slightly blurring it and encouraging use of the slightly amblyopic eye while still yielding fusion.

Treatment of diplopia by occlusion requires occlusion of either eye depending on which eye the patient prefers to have occluded. (Occlusion amblyopia is not a factor for older patients being occluded in this way.) The occlusion either directly eliminates the diplopia and/or helps prevent permanent muscle contractures.

The actual occlusion may be done in one of several ways and patients should be treated individually in this respect. An on-face occlusive patch may be used, but generally is bothersome to wear for an extended period of time. The black "pirate patch" is more comfortable but is cosmetically obvious. Other patients will prefer occluding the eyeglass lens either with opaque tape or with transparent tape, nail polish, clear contact paper, or Bangerter filters. If this occludes them adequately, it also gives patients better cosmesis as the occlusion is less obvious to the public. A high plus spherical press-on Fresnel lens may also be used to occlude.

Some patients are still aware of diplopia even with CF visual acuity. High plus over-correction contact lenses may help those who can afford lenses, but again, it may not occlude them adequately. Opaque iris lenses (prosthesis) may be a last resort. Appropriately dissuade the patient who wants enucleation as a permanent treatment of diplopia.

Because suppression and ARC are *binocular* phenomena, occlusion removes the young patient's binocularity and therefore the need to suppress or have ARC. Direct on-face occlusion works best for these patients, but young patients must be checked frequently in order to avoid occlusion amblyopia and to monitor the progress. Alternate eye

Methods of Occlusion
1. On-face patch
2. "Pirate patch"
3. Nonopaque occlusion: transparent tape, nail polish, clear contact paper, Bangerter filter
4. High plus Fresnel spherical lens
5. High plus contact lens
6. Opaque iris lens

occlusion is used if the patient has absolutely equal vision. More occlusion may be done on the preferred eye of a young amblyope.

In patients with fusion potential, orthoptic treatment is done to improve their fusion or to make them aware of diplopia. Diplopia awareness is done most effectively with cooperative patients who are willing to carry out the exercises on a daily basis at home. By working with patients to improve their resistance to obstacles to fusion, their fusional control is improved. Vergence amplitudes must be adequate for the patient's needs. Fusional convergence is most effectively taught using stereograms (see Figure 8.1) or with hand held base-out prisms for reading (see Figure 8.2).

Pharmacologic

Drugs are used to treat two types of ocular motility problems. *Penalization* involves using drugs and glasses to force use of one eye or the other in the treatment of amblyopia after traditional on-face occlusion (see Chapter 4).

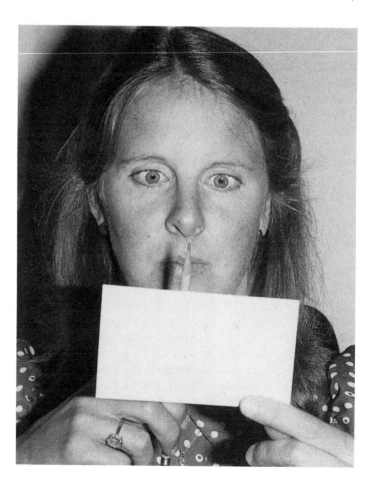

Figure 8.1 Patient converging on pencil using physiologic diplopia to fuse stereogram card and increase fusional convergence.

Anticholinesterase inhibitors such as *phospholine iodide* (PI) or *diisopropyl fluorophosphate* (DEP) are used in strabismus therapy to reduce the ET that is due to a high AC/A ratio. While spectacle correction invariably works better (without the side effects of the drugs), in some cases PI is a good trial to determine if an ET is actually due to a high AC/A ratio. It is particularly useful to use PI when the child absolutely cannot wear glasses, or when it is not certain if glasses will actually straighten the ET, and the expense of the glasses to the parents should be spared.

Side effects of PI/DFP
1. Succinylcholine reaction
2. Iris cysts
**All children using PI/DFP should wear a medical alert bracelet.

Glasses

Spectacle correction is used to treat refractive errors and is particularly important in young children where amblyopia may develop or where poor vision is an obstacle to fusion. To diminish this obstacle to fusion, a *balanced correction* is given to allow both eyes to see clearly at the same time with the same amount of accommodation. Correction of high cylinders is also necessary to prevent *meridional amblyopia*.

Glasses treat:
1. Refractive errors
2. Accommodative ETs

Figure 8.2 Patient reading with base-out prism in order to increase fusional convergence amplitudes.

Glasses are also used in children in the treatment of accommodative ET with or without high AC/A ratios. These children are usually hyperopic. When the children accommodate in order to see clearly, their associated accommodative convergence cannot be balanced by divergence resulting in ET. Full correction of the hyperopia results in no need for accommodation and therefore no esotropia due to accommodation.

Some children have a particularly high AC/A ratio that causes their eye to cross in even more when they fixate on a near object. This near deviation is eliminated with the use of a bifocal add. Except in cases where there is high hyperopia that would require correction so that the child can see clearly, accommodative ETs with or without high AC/A ratios require glasses only to keep the eyes straight and fusing. These children do not require the glasses to see clearly. This distinction is very important to point out to parents so that they understand why the child is initially resistant to wearing the glasses. The glasses simply do not make the child see better, and frequently, until the child relaxes his accommodation, the glasses may make him see worse.

Initially, the full amount of hyperopia is prescribed and the plus add that is required for fusion is given. Once fusion is regained by the patient, the *least* amount of plus that still yields fusion is given to the patient.

The segment height of an add for a child with accommodative ET is important for successful treatment. In contrast to an adult where you do not want the line to interfere with the straight ahead distance viewing, the bifocal segment in a child must be high enough to bisect the pupil with the eyes looking straight ahead (see Figure 8.3). The child must be

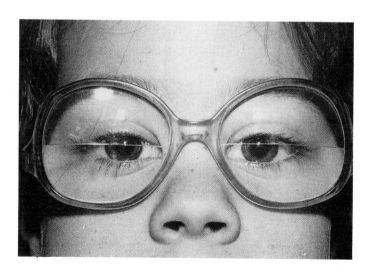

Figure 8.3 Executive bifocal in treatment of accommodative ET with high AC/A ratio. Segment height bisects pupil so that child is encouraged to use bifocal for near fixation.

forced into using the bifocal when viewing near objects. This instruction must be put onto the prescription form because many opticians do not routinely fill such prescriptions for small children and would not automatically do it.

Ordinarily, an executive type bifocal that goes all the way across the bottom of the spectacle glass would be prescribed so that the child cannot look around the segment when looking down. Executive bifocals are heavy and expensive and in a small child's frame, a "D-35" segment can be safely substituted. Because the bifocal is 35 mm in width, and the child's frame is not much wider than that, a child who is familiar with the use of the bifocal will not peek around the "D-35" much (see Figure 8.4).

Press-on Fresnel lenses may also be used as temporary bifocal adds. These are particularly helpful when it is not certain if the additional adds will straighten out the deviation and the parents want to be spared the expense of the bifocal glasses. When a change in the glasses is necessary shortly after the original glasses have been obtained, the press-on lenses can be used to increase or decrease the power of the correction.

A patient who changes from hyperopic spectacles to contact lenses correction will experience a decreased need to accommodate at near and therefore the accommodative ET will benefit. The hyperopic *exo* patient will get less natural accommodative convergence with the contacts than with the glasses and therefore may need to exert *more* fusional convergence (Rubin, 1974).

A patient who changes from myopic spectacles to contact lenses will experience an *increased* need to accommodate and therefore will have more accommodative convergence (good news for an exo, bad news for an eso).

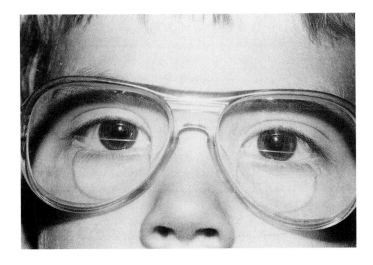

Figure 8.4 "D-35" type bifocal in treatment of accommodative ET with high AC/A ratio.

Prism Therapy

Prisms need a pair of glasses to be hung onto whether they are made directly into the glasses or are dispensed as a Fresnel press-on prism.

Constant diplopia may be treated with small amounts of prism, particularly if the deviation is comitant. Intermittent diplopia may also be treated occasionally with prism but only if fusional vergences cannot be improved.

Often a major error is made when patients with small fusible exodeviations are given base in prism in their spectacles. Of course it helps them initially, but the long-term effects are undesirable. The base in prism allows the patient's eyes to relax into an exotropic position underneath the glasses. The eyes continue to fuse, but are allowed to exert less fusional convergence. The prisms act as a crutch, thereby reducing the need to exert motor fusion. Eventually this crutch leads to the patient having less motor fusion, so the patient will need more base in prism. This is the so-called, "eating up the prism." The process may take years, but the end result is usually the same: Eventually the patient is constantly diplopic without the prism and is committed to wearing a cosmetically and visually unacceptable amount of prism. The patient is then either committed to surgery or to fusional convergence exercises. If the patient wants to avoid surgery, he will diligently do convergence exercises until he can be weaned off of the base in prism until no prism is necessary. The best way to treat these patients *initially*, is to improve their fusional convergence.

The one exception to this approach is with the elderly patient. A patient more than seventy years old does not necessarily have to worry about how much prism he will be using in another twenty-five years. Elderly patients will small exodeviations are best treated with the minimum amount of base in prism that allows fusion.

When vertical prism is used, base UP prism should be given when possible. This allows the patient to read in down gaze through the thinnest part of the prism, the apex.

Combined horizontal and vertical deviations may be corrected with obliquely placed prism. Trial and error with press-on prisms is the most economical way of doing this. The appropriate power and axis can be determined mathematically if the horizontal and vertical prism powers are known. (Moore, Stockbridge, 1972) It is also helpful to experiment to determine if only the horizontal and vertical deviation needs to be corrected. A patient with excellent horizontal fusional control may be able to fuse successfully with *only* the vertical deviation corrected.

Prismatic glasses treat:
1. Some types of diplopia
2. Head positions due to nystagmus

When a patient requires a completely different amount of prism at distance and near, or if prism is necessary *only* at distance or near, press-on prisms must be used. They can be cut out in pieces to be used at distance or near.

The head position that a patient with nystagmus uses may be cosmetically unacceptable. Prisms may be used to eliminate the *head position* by allowing the eyes to deviate underneath the prisms. For example, a patient with a head position to the right uses left gaze. With the head straight, prisms base RIGHT (BO OD, BI OS) will force the eyes into left gaze and the null point. The head will be straight; the eyes will be permanently in left gaze underneath the prisms.

Nystagmus also dampens when the patients converge. Some patient's vision will improve with base out prisms or over-minused lenses that cause accommodative convergence to be exerted.

Surgery

Surgical correction of strabismus is sometimes used as a last resort and other times used as a primary treatment. If the patient has fusion potential that cannot be restored successfully with other treatment, surgery is performed to restore comfortable fusion. Surgery may be indicated in the following cases: adult onset cranial nerve palsies, intermittent exodeviations, Grave's ophthalmopathy, or the need to eliminate the head position in patients with nystagmus.

EOM Surgery is done to:
1. Restore fusion
2. Improve cosmesis
3. Increase ocular movement

In the absence of fusion potential, surgery is done to improve cosmesis. Some patients, such as those with infantile ET and XT, or DVD develop rudimentary peripheral fusion after surgery. The possibility of postoperative diplopia must be considered when cosmetic surgery is suggested for an adult. Patients with CF vision in one eye may be tremendously bothered by diplopia if the eye position is moved from what they are accustomed to. A patient with a very large deviation and diplopia may be more bothered with a much smaller cosmetically acceptable deviation if the diplopic images have been moved closer together.

One general rule does apply to strabismus surgery: Every patient is different and no *firm* rules apply. Weakening procedures (recessions, myotomies, posterior fixation sutures) of normal healthy muscles generally do work better than strengthening procedures (resections, tucks) of weak palsied muscles. Ophthalmologists establish their own guidelines for how much surgery corrects how much deviation (Helveston, 1977).

An incomitant deviation will probably be slightly incomitant postoperatively. The patient's eye turn in the primary position, and reading position if they fuse, is considered

Table 8.1 Duane's Classification of Horizontal Strabismus

Deviation	Surgical choice
Basic deviation	R + R on nonpreferred eye
Convergence excess	Bilateral MR recession
Convergence insufficiency	Bilateral MR resection
Divergence excess	Bilateral LR recession
Divergence insufficiency	Bilateral LR recession
Pseudo divergence excess	R + R okay

Table 8.2 A and V Patterns

A patterns	Surgical choice
Due to overactive: SR, SO	Weakening procedures
Due to underactive: MR, LR, IR, IO	Strengthening procedures
No EOM dysfunction	Move MR up, LR down (**MALB)

V patterns	Surgical choice
Due to overactive: MR, LR, IR, IO	Weakening procedures
Due to underactive: SR, SO	Strengthening procedures
No EOM dysfunction	Move MR down, LR up (**MALB)

**MALB "Medials to the Apex, Laterals to the Base"

MALB is a mnemonic to remember which way the horizontal muscles are surgically moved to correct an A or V pattern when there is no EOM dysfunction. Consider an "A" to be like a prism base down apex up, and a "V" to be like a prism base up apex down. MALB (**M**edials to the **A**pex, **L**aterals to the **B**ase) means that in an **A pattern**, the MR would be moved **up** (apex) and/or the LR would be moved **down** (base). In a **V pattern**, the MR would be moved **down** (apex) and/or the LR would be moved **up** (base).

most important to surgically correct. The patient's eye position in extreme upgaze, on the other hand, is of little importance unless the patient is routinely looking for enemy aircraft.

Some guidelines will help reduce the incomitant nature of the deviation. Duane's classification of horizontal strabismus (see Chapter 3) recognizes that the MR are mainly responsible for near deviations, and the LR are mainly responsible for distance deviations. (see Table 8.1).

A and V patterns may be due to over- or underactions of EOMs and when the appropriate muscle at fault is determined, it should be addressed directly (see Table 8.2).

After the spread of comitance has occurred following a cyclovertical muscle palsy, muscles may be weakened or strengthened depending on what field of gaze is to be restored.

References

Helveston, E.M. *Atlas of Strabismus Surgery,* 2nd ed. Saint Louis: C.V. Mosby, 1977, p 210.

Moore, S. Stockbridge, L. Symposium: The Use of Prisms in the Management of Nonparetic Strabismus—Fresnel Prisms in the Management of Combined Horizontal and Vertical Strabismus. *American Orthoptic Journal,* 22:14-21, 1972.

Rubin, M.L. *Optics for Clinicians.* 2nd ed. Gainesville, Fla: Triad Scientific Publishers, 1974, p 267.

CHAPTER 9

Prognosis—
The OMA's Role

General Factors That Influence Prognosis

Age of Onset

A child with a motility problem that started at birth is likely to have developed suppression, amblyopia, and/or ARC. Congenital findings are usually diagnosed and treated early. When the age of onset of the motility problem is 6 years of age or older, the chance of developing amblyopia is greatly reduced. Suppression and/or ARC can develop up to an age of onset of 15 years old. On the other hand, an adult who develops a permanent strabismus problem with a loss of binocular fusion, may never learn to suppress and will therefore be permanently bothered by diplopia.

Age of Treatment Initiation

The prognosis of a patient with an ocular motility problem improves when the appropriate treatment is initiated at a young age soon after the diagnosis is made. The chances of success of the treatment are diminished when time is wasted between the diagnosis and the beginning of treatment. Early treatment requires early diagnosis which is one reason why early recognition of ophthalmic diseases by pediatricians is very important. Education of pediatricians by ophthalmologists is some communities may be helpful.

General Factors Influencing Prognosis
1. Age of onset
2. Age of treatment initiation
3. Follow-up appointment compliance
4. Treatment compliance
5. Depth of adaptations to strabismus at first visit

Follow-up Appointment Compliance

Once treatment is initiated, its success or failure must be frequently monitored. Patients cannot treat themselves. Patients are more likely to keep their follow-up appointments when they understand the reasons for frequent visits and the potential results of their negligence. Parents need to know that many strabismic and amblyopic children will require years of follow-up care.

Treatment Compliance

When patients understand the purpose of their treatment and the consequences of their noncompliance, they are more likely to do what they are told. Parents must understand that they are responsible for the outcome of their child's visual status and that the proper treatment can successfully correct the problem. Children cannot make rational decisions about the future of their eyes.

Depth of Adaptation to Strabismus at First Visit

The prognosis improves when a child's adaptations are mild and superficial. The closer the adaptations are to normal, the better the prognosis is. A 20/200 amblyope is less likely to reach 20/20 than someone starting with 20/50 vision. A patient with a well-established ARC and suppression pattern is less likely to obtain fusion than someone else with superficial ARC or suppression.

Of course, all of these factors must be balanced against each other and then the prognosis can be given to the parents. Remember that every child, every case, and every parent is different making a 100% accurate prognosis somewhat difficult. Usually, after a few follow-up visits, there is a better idea about how good the treatment compliance is and how the child is responding to the therapy. An updated prognosis can then be made.

Individual Factors That Influence Prognosis

Other Factors Influencing Prognosis:
1. Use of head position for fusion
2. Fixation pattern
3. Fusion potential
4. Diplopia awareness

Additional individual factors will help determine if a child will successfully reach the best visual potential possible. For instance, a young child who has a *head position* is probably trying to maintain fusion. These children continue to fuse when surgery straightens the ocular motility problem that caused the head position.

A young child who *alternates* fixation is less likely to be amblyopic and is also less likely to require strict maintenance patching once equal vision is achieved.

Fusion potential is a major factor when a patient is being considered for surgery and dramatically improves the chances of a perfect postoperative eye position with fusion.

Diplopia awareness enhances the patient's awareness of his eye turn and signals him when to use motor fusion to achieve sensory fusion.

What Can the Parents Expect?

Amblyopia

Occlusion therapy that starts on a three-year-old child will initially be full time at least for several months and will require frequent follow-up appointments (more appointments for a younger child). Once equal vision is achieved, partial occlusion will be necessary for a number of years, but with less frequent follow-up appointments. The children must be followed at least until age 8 but sometimes up to age 15. When amblyopia is unsuccessfully treated, either because of poor patient/parent compliance or because the treatment was started too late, precautions must be taken to protect the non amblyopic eye from the unexpected loss of vision.

Amblyopic children should wear protective eye-wear that is appropriate for their activity and should have routine eye exams throughout their lives so as to detect and prevent other eye disease. (Vinger, 1982)

Surgery

People still have wrong ideas about what strabismus surgery involves. And many people simply do not want to know what is done during surgery. Usually it is helpful to ask the patient if there are any questions, and if there are none, tell the patient a little about exactly what happens. For instance, say: "No, they don't take the eye out"; "There will be no more scars on the skin but it may be the reddest eye you've ever seen." The conjunctival injection disappears rapidly and it looks much worse than it feels to the patient. The patient will not be able to "see" the operation. Activities postoperatively should be restricted for several days; no recess, gym, or contact sports. Activities where the patient is likely to be pushed or poked should be avoided. And finally, tell parents that it is much worse for *them* than for their child having the surgery! It really is. Older patients may expect to see double for a few days to weeks afterwards, particularly if they were exo pre-op and an over-correction is planned. In a certain percentage of patients, a second surgery will be necessary and in infantile ETs, a DVD often does not become obvious until after the horizontal deviation has been corrected.

Fusion Potential

The ultimate goal of therapy is comfortable single binocular vision that yields fusion; patients with good fusion potential are most likely to be able to attain this. Patients who have had a constant eye turn since birth, and/or amblyopia are much less likely to develop fusion. At best, they may develop peripheral fusion. The consequences for a child who never develops excellent bifoveal fusion are minimal.

Inform Parents Regarding:
1. *Amblyopia*
2. *Surgery*
3. *Fusion potential*
4. *Diplopia and ARC*

First, children who never had fusion or stereopsis don't miss it. These children can go up and down stairs, pour milk, or parallel park their tricycle just as easily an anyone else. Stereopsis is most "missed" when viewing small fast moving objects, such as a tennis ball, or when trying to catch a fly baseball against a clear blue sky where there are no peripheral clues to depth perception. Bifoveal stereopsis makes it easier to do these things but is not necessarily going to make or break a career in professional sports. (Wesley Walker, receiver for the NY Jets football team has had a blind eye since childhood without any true stereopsis.)

When parents of young children without stereopsis earnestly ask about the consequences, tell them honestly. Monocular pilots will not be issued a commercial airline pilot's license and some individual public transportation companies have rules about their drivers having stereopsis. "3-D" movies won't be any fun. Some medical schools or special residency programs have binocular vision requirements, but many do not. Tell the parents that if the child seems to like basketball as much as baseball, to *maybe* encourage basketball, since the ball is bigger and moves slower. Other than that, the child is not "handicapped."

Diplopia and ARC

An older patient with deeply established ARC who has surgery for cosmetic reasons runs the risk of having permanent postoperative diplopia (see Chapter 4). Predicting who will be diplopic helps the patient decide if surgery is worth it or if only part of the deviation can safely be operated upon without the risk of postoperative diplopia.

Summary

Happy patients are those patients who got what they expected and were also given the best possible treatment. The OMA can play a critical role by explaining the expectations, particularly for patients who have ocular motility problems.

References

Vinger, P. Preventing ocular injuries. *American Orthoptic Journal* 32:56–60, 1982.

Index